Putnam's Elementary Science Series.

INORGANIC CHEMISTRY:

FOR USE IN

SCIENCE CLASSES AND HIGHER AND MIDDLE CLASS SCHOOLS.

BY

DR. W. B. KEMSHEAD, F.R.A.S., F.G.S.,

LECTURER ON CHEMISTRY AND NATURAL SCIENCE, DULWICH COLLEGE, LONDON; AND EXAMINER IN CHEMISTRY, COLLEGE OF PRECEPTORS, LONDON.

NEW YORK:
G. P. PUTNAM'S SONS,
FOURTH AVENUE AND TWENTY-THIRD STREET.

TO THE

Rev. A. J. Carver, D.D.,

Master of Dulwich College,

THIS LITTLE WORK IS
RESPECTFULLY INSCRIBED, AS A SLIGHT RECOGNITION
OF PAST KINDNESSES AND COURTESIES
RECEIVED BY

THE AUTHOR.

PREFACE.

The following work does not pretend to be a complete text-book on the science of Inorganic Chemistry, or even on that portion of it which would confine itself to the Non-metals. Written with an especial purpose, viz., for the use of pupils preparing for the First Stage, or Elementary Examination of the Science and Art Department, South Kensington, it necessarily confines itself to the subjects prescribed in the Syllabus of that Examination.

My aim throughout has been to express, in as clear and simple language as possible, the earlier principles of the science, so as to fit the book for the use of mere beginners, but at the same time to be sufficiently full and accurate that it might be useful as a text-book in the hands of more advanced students. The limits to which this series is confined have prevented me from giving all the experimental illustrations I could have wished; but I have, I trust, given amply sufficient to illustrate every assertion in the text, and have sought to make choice of those which, while they are striking and conclusive, combine also the property of being easily performed, and are therefore most suitable to students whose command of apparatus may be limited.

I make no apology for introducing the Graphic Formulæ and Notation in so elementary a work. Independently of my own proclivities in its favour, the examination for which this book is specially prepared demands a knowledge of the theory of atomicity and of its graphic representation—to have omitted it, therefore, would have been impossible. For the benefit of those teachers who may not have adopted it, I have in all cases added the equations in the ordinary symbols as used by the late Professor Miller, Professor Williamson, &c.

I have to offer my warm acknowledgments to Professor Frankland for the ready and kind manner in which he gave me permission to make free use of his valuable work, *Lecture Notes for Chemical Students.*

The elementary character of the work did not admit of much originality, except perhaps in the matter of arrangement; and I must here acknowledge my obligations to the works of Faraday, Miller, Williamson, Bloxam, &c., &c.

The work while in the form of notes has done good service in preparing my own pupils for the South Kensington Examination; I trust it may be equally successful in the hands of those teachers who may adopt it.

W. B. KEMSHEAD.

DULWICH COLLEGE, LABORATORY,
DULWICH, *September, 1873.*

CONTENTS.

CHAPTER I.

Matter and Force—Physical Forces—Definition of Chemistry—Simple and Compound Matter—Metals and Metalloids or Non-Metals—Symbols—Atomic Weights—List of Elements—Physical Condition of Elements—Distribution of Elements, ... Pages 9-16.

CHAPTER II.

Difference between Mechanical Mixture and Chemical Compound—Characteristics and Different Modes of Chemical Action—Summary, ... Pages 16-26.

CHAPTER III.

Combining Weights—Laws of Chemical Combination—Outline of Atomic Theory—Combining Volumes—Volume Weights, Pages 26-38.

CHAPTER IV.

French and English Systems of Weights and Measures—Conversion of English into French Weights and Measures—The Crith and its Uses, ... Pages 38-43.

CHAPTER V.

Principles of Chemical Nomenclature—Classification of Elements into Positive and Negative—Symbolic Notation—Chemical Formulæ—Chemical Equations, ... Pages 43-53.

CHAPTER VI.

Atomicity of Elements—Classification according to Atomicity—Graphic Notation—Simple and Compound Radicals—Definition of a Compound Radical, ... Pages 53-65.

CHAPTER VII.

Hydrogen—Its History—Distribution and Natural History—Preparation—Properties—Combinations, . Pages 66-82.

CHAPTER VIII.

Chlorine—History—Distribution and Natural History—Preparation—Properties—Combinations, . . Pages 82-90.

CHAPTER IX.

Hydrochloric Acid—History—Distribution and Natural History—Preparation—Properties—Combinations, . Pages 90-98.

CHAPTER X.

Oxygen—History—Distribution and Natural History—Preparation—Properties—Allotropic Oxygen or Ozone, Pages 98-111.

CHAPTER XI.

Combinations of Oxygen—Formation and Reactions of Water—Preparation and Properties of Hydroxyl—Compounds of Chlorine with Oxygen and Hydroxyl, . Pages 112-129.

CHAPTER XII.

Boron—History—Its Occurrence in Nature—Its Allotropic Modifications—Boric Anhydride—Boric Acids, . Pages 129-133.

CHAPTER XIII.

Carbon—Preparation—Allotropic Forms—Combinations—Preparation and Properties of Carbonic Oxide—Carbonic Anhydride, its Preparation and Properties, . Pages 133-145.

CHAPTER XIV.

Nitrogen—Its Preparation and Properties—Compounds of Nitrogen with Oxygen and Hydrogen—Compounds of Nitrogen with Hydrogen—Ammonia—Ammonic Salts, Pages 145-156.

CHAPTER XV.

Compounds of Nitrogen with Hydrogen—Ammonia—Its History, Preparation, and Properties — Ammonium — Ammonic Salts, ... Pages 156-163

CHAPTER XVI.

Sulphur—History—Occurrence—Preparation—Properties—Allotropic Modifications—Uses—Compounds of Sulphur with Positive Elements—Sulphuretted Hydrogen—Occurrence—Preparation—Properties—Reactions—Uses—Hydrosulphyl—Carbonic Disulphide—Properties—Reactions, Pages 163-169.

CHAPTER XVII.

Compounds of Sulphur with Oxygen and Hydroxyl, Pages 169-183.

INORGANIC CHEMISTRY.

CHAPTER I.

Matter and Force—Physical Forces—Definition of Chemistry—Simple and Compound Matter—Metals and Metalloids or Non-Metals—Symbols—Atomic Weights—List of Elements—Physical Condition of Elements—Distribution of Elements.

WE shall the better understand the full definition of the science of **Chemistry,** if we first explain some of the terms which we are compelled to make use of.

Of these terms the two most important for us clearly to understand are **matter** and **force.**

1. **Matter** is the name which we give to all *things* which exist, or which can in any way be recognized by our senses; thus, the earth we live on, the water we drink, the air we breathe, the bodies we dwell in, are all material, or, scientifically speaking, masses of *matter.*

Matter possesses various properties, but the essential ones (without which, in fact, it would not be *matter*) are, that it must possess weight and occupy space.

All *matter* is subjected to various influences, some of which impart to it new properties, without, however, changing its external form or appearance; others, while altering its external form, do not sensibly change it in other respects; while another, that with which we are more directly concerned, not only alters matter in external form, but in nearly all cases imparts to it totally new and different properties.

These various influences are called **forces,** and are divided into **physical or natural and chemical forces—**

e. g., if a bar of iron or any metal be heated, it will become visibly, sensibly larger; and if the heat be sufficiently great, it will pass from the solid to the liquid form, but it will still remain iron, or copper, or brass, &c., as the case may be. *Heat*, therefore, simply changes its size and condition, but not its nature.

EXPERIMENT 1.—Thus, if some water be put into a Florence flask, and heat be applied by means of a spirit lamp or Bunsen burner, the water will be converted into vapour which will be invisible; but if an arrangement similar to that shown in Fig. 1, where A is a retort in which the water is heated; B, a receiver, kept cool

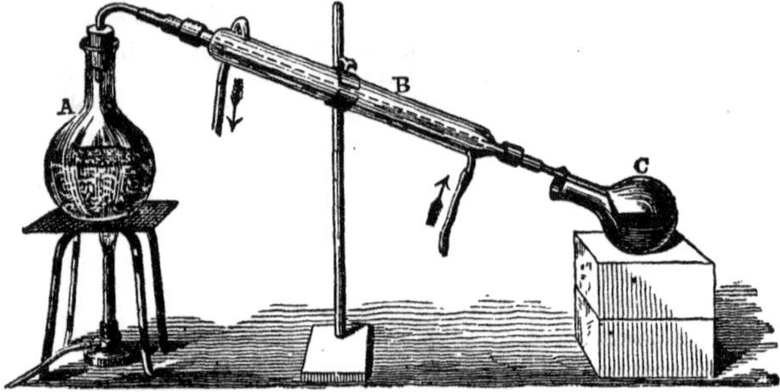

Fig. 1.

by a constant supply of cold water kept up over the outside; and C, a vessel to receive the water thus condensed, the vapour will be again converted into water, from which it started. If now some water be put in a metal vessel—a tin, or zinc, or pewter pot, and the whole then immersed in a mixture of broken ice and salt, the water will gradually contract or diminish in volume, until it reaches a temperature a little above freezing point (4° C. or 39·2° F.), when it expands, and continues to do so until the moment of solidification, when it expands suddenly and violently to $\frac{1}{12}$ of its former bulk, and finally it will be converted into a solid mass of ice. But whether it be in the condition of vapour or gas, the liquid water, or the solid ice, it is still the same chemical substance—*water* (OH_2).

EXP. 2.—If some *iodine* be put in a Florence flask, and a gentle heat applied, the iodine will not melt, but will at once be converted into a beautiful rose-coloured vapour, which will be again condensed into minute crystals of iodine in the upper or cool part of the flask.

These experiments show that *heat* may alter the size and physical condition of a substance; but, especially if the substance be a simple one, it does not alter its nature.

Again, if a rod of sealing wax or glass be rubbed with a piece of warm dry flannel or silk, they will undergo no apparent change of any kind, but they will be found to have acquired the property of attracting to themselves all light substances. This property we call **electricity.**

Further, if a bar of steel or a steel needle be rubbed with a piece of loadstone, it will present the same appearance; but if suspended, so as to be free to move, it will now always point in one direction—it is, in fact, under the force of **magnetism.**

But if we put together some iron or copper filings and sulphur, and expose the mixture to heat, the metal and the sulphur will combine, and a third substance will be formed which will be found to possess properties totally distinct from those of the metal or sulphur employed. The *force* under which this combination takes place is called **chemical force.**

2. Chemistry, then, is the science which treats of *matter* (*i. e.*, of all bodies whatsoever which have been or can be subjected to man's investigation) in respect of its nature, its composition, and its properties. It investigates the mutual relations of different masses of *matter*, and their actions on each other when under the influence of this *chemical force*.

3. Simple and Compound Matter.—One of the first properties of matter which forces itself on the attention of the chemist is, that it is either *simple* or *compound* in its nature.

4. By **simple bodies** or **elements** we are to understand those which we are not able by any means we have at our command to split up into two or more different substances. Thus, *sulphur* is an *element;* by no process whatever can we obtain anything but sulphur from it. *Mercury*, *copper*, *iron*, and *gold* are all *simple substances* or *elements*.

5. By **compound bodies** we mean all those which, under the treatment of the chemist, can be made to yield more than one simple or elementary body—*e. g.*, chalk is a compound; under suitable treatment it can be made to yield an invisible gas called oxygen, a black solid—carbon, and a white shining metal—calcium. So, sugar is a compound, and may be resolved into two gases—oxygen and hydrogen, and a solid—carbon. Brass is a compound, and may be divided into the two metals, zinc and copper.

NOTE.—All compounds formed by the union of two or more metals are called *alloys*, unless mercury be one, in which case they are called *amalgams*.

Of these *elements* chemists have ascertained the existence of sixty-two or sixty-three. These have been again subdivided into two classes—**non-metals** or **metalloids,** and **metals.**

NOTE.—The former term, although not so convenient, because a compound word, is more correct, and is now universally employed.

6. Characteristics of Metals and Non-Metals.—The *metals* are generally known as possessing a peculiar lustre called metallic, a high specific gravity, or great density, and for being good conductors of heat and electricity; while the *non-metals* are noted for the absence of this metallic lustre, their generally low specific gravity, and for their being generally bad conductors of heat and electricity.

The above definition, though perhaps the best that can at present be given, is not at all satisfactory; for, with respect to lustre, many of the metals possess it only in a modified degree, and some lose it altogether when reduced to powder; while one of the non-metals, *iodine*, possesses it in a very remarkable degree. Again, with respect to density or specific gravity, while it is true that the metals for the most part are heavy, while the metalloids are light, it is not universally the case; for three of the metals—*lithium, potassium,* and *sodium*—are lighter than water; and several of the *non-metals*—bromine, selenium, iodine, arsenic—have a density from three to five times that of water.

In like manner, the power of conducting *heat* and electricity, although distinctive of, is not absolutely confined to, metals.

7. Symbols.—Instead of writing the whole name of an element, chemists are in the habit of using merely the initial letter, or, in the case of two or more elements beginning with the same, the more common or more important one has the initial; while the other elements are distinguished by the addition of a small letter, thus—*carbon* (C), *calcium* (Ca), chlorine (Cl), cobalt (Co), cæsium (Cs), and so on.

8. Atomic Weights.—It is found by experiment that whenever substances, either elementary or compound, unite together chemically, they *always* do so in certain fixed definite proportions. These proportions by weight, when reduced to their lowest relative value, and expressed with reference to that of *hydrogen*, which is usually taken as unity, are called the **atomic weights,** or **combining numbers,** of the elements. Thus, the **symbols** not only indicate the elements which they represent, but they also stand for a fixed definite proportion by weight of that element. Subjoined is a list of the elements, with their symbols and atomic weights, or combining numbers:

Names of Elements.	Symbols.	Atomic Weights.
Aluminium,	Al.	27·5
Antimony (Stibium),	Sb.	122
Arsenic,	As.	75
Barium,	Ba.	137
Bismuth,	Bi.	210
Boron,	B.	11
Bromine,	Br.	80
Cadmium,	Cd.	112
Cæsium,	Cs.	133
Calcium,	Ca.	40
Carbon,	C.	12
Cerium,	Ce.	92
Chlorine,	Cl.	35·5
Chromium,	Cr.	52·5
Cobalt,	Co.	59
Copper (Cuprum),	Cu.	63·5
Didymium,	Di.	95

Names of Elements.	Symbols.	Atomic Weights.
Erbium, ...	E.	112·6
Fluorine, ...	F.	19
Glucinum or Beryllium, ...	G. or Be.	9·3 (?)
Gold (Aurum)	Au.	196·6
Hydrogen, ...	H.	1·
Indium, ...	In.	113·4
Iodine, ...	I.	127
Iridium, ...	Ir.	197
Iron (Ferrum), ...	Fe.	56
Lanthanum, ...	La.	92
Lead (Plumbum), ...	Pb.	207
Lithium, ...	Li.	7
Magnesium, ...	Mg.	24
Manganese, ...	Mn.	55
Mercury (Hydrargyrus), ...	Hg.	200
Molybdenum, ...	Mo.	96
Nickel, ...	Ni.	59
Niobium or Columbium, ...	Nb.	97·5
Nitrogen, ...	N.	14
Osmium, ...	Os.	199
Oxygen, ...	O.	16
Palladium, ...	Pd.	106·5
Phosphorus, ...	P.	31
Platinum, ...	Pt.	197·4
Potassium (Kalium), ...	K.	39
Rhodium, ...	Rh. or Ro.	104
Rubidium, ...	Rb.	85·5
Ruthenium, ...	Ru.	104
Selenium, ...	Se	79
Silicon or Silicium, ...	Si.	28·5
Silver (Argentum), ...	Ag.	108
Sodium (Natrium), ...	Na.	23
Strontium, ...	Sr.	87·5
Sulphur, ...	S.	32
Tantalum, ...	Ta.	137·5
Tellurium, ...	Te.	128
Thallium, ...	Tl.	204
Thorinum, ...	Th.	231·5
Tin (Stannum), ...	Sn.	118
Titanium, ...	Ti.	50
Tungsten (Wolfram), ...	W.	184
Uranium, ...	U.	120
Vanadium, ...	V.	137
Yttrium, ...	Y.	68
Zinc, ...	Zn.	65
Zirconium, ...	Zr	89·5

9. Non-Metals.—The following *thirteen elements* are universally admitted as **non-metals**:—

Hydrogen, ...	H.	1	*Boron,*	B.	11
Oxygen, ...	O.	16	*Iodine,*	I.	127
Nitrogen, ...	N.	14	*Phosphorus,*	P.	31
Chlorine, ...	Cl.	35·5	*Sulphur,*	S.	32
Fluorine, ...	F.	19	*Selenium,*	Se.	79
Bromine, ...	Br.	80	*Silicon,*	Si.	28·5
Carbon, ...	C.	12			

To these the majority of chemists add *tellurium* (Te, 128) and *arsenic* (As, 75). The remainder of the *elements* are **metals.**

10. Physical Condition of Elements.—Under the ordinary circumstances of temperature and pressure, the following five *elements* are *gaseous,* viz.; oxygen (O), hydrogen (H), nitrogen (N), chlorine (Cl), and fluorine (F); while bromine (Br) and mercury (Hg) are *liquid,* and the remainder are all *solids.*

11. Distribution of Elements.—The following twelve elements constitute the chief part of the earth, whether of the solid ground, the sea, the air, or the animals and vegetables that inhabit them; they are consequently the most abundant:—

Hydrogen, ...	H.	*Silicon,*	Si.
Oxygen, ...	O.	*Aluminium,*	Al.
Nitrogen, ...	N.	*Calcium,*	Ca.
Carbon, ...	C.	*Iron,*	Fe.
Chlorine, ...	Cl.	*Potassium,*	K.
Sulphur, ...	S.	*Sodium,*	Na.

The next eleven elements, although not so abundant, are either of frequent occurrence, or of great chemical importance. They are:—

Bromine, ...	Br.	*Manganese,*	Mn.
Copper, ...	Cu.	*Mercury,*	Hg.
Fluorine, ...	F.	*Phosphorus,*	P.
Iodine, ...	I.	*Silver,*	Ag.
Lead, ...	Pb.	*Zinc,*	Zn.
Magnesium, ...	Mg.		

The next group of eighteen elements may be regarded as of secondary importance, viz.:—

Antimony,.............	Sb.	Nickel, ...	Ni.
Arsenic,................	As.	Palladium,.............	Pd.
Barium,................	Ba.	Platinum,..............	Pt.
Bismuth, ...	Bi.	Rhodium, ...	Rh.
Boron,..................	B.	Strontium,.............	Sr.
Chromium, ...	Cr.	Tin,......................	Sn.
Cobalt, ...	Co.	Titanium, ...	Ti.
Gold,....................	Au.	Tungsten, ...	W.
Iridium,................	Ir.	Uranium, ...	U.

The remaining elements may be regarded as those of very rare occurrence, or of which our knowledge is yet very imperfect. Some of them, indeed, as erbium, indium, &c., are at present merely chemical curiosities.

CHAPTER II.

Difference between Mechanical Mixture and Chemical Compound —Characteristics and Different Modes of Chemical Action —Summary.

12. We have defined "a compound body" to be one which we can decompose into two or more simple bodies or elements; but in considering this definition it is important to distinguish clearly between a mere *mechanical mixture* and a *chemical compound*, for the effects produced by mixture and by combination are exceedingly different.

13. Properties of a Mechanical Mixture.—In a mixture, the materials may exist in any proportion whatever, and the properties of the mixture will partake of those of each of its constituents; while in a chemical compound combination will only take place in certain fixed, definite, and unalterable proportions, and the resulting compound will possess properties totally distinct from those of either of its constituents.

Exp. 3.—If iron filings and powdered sulphur be mixed together, they may be so in any proportion whatever, and the resulting mixture will retain in a modified form the characteristics of both. Thus the inflammability of the sulphur will be modified by the non-inflammability of the iron, and the power of conducting electricity which the iron possesses will be modified by the presence of the sulphur, which does not possess this power; but the iron and the sulphur will still exist independent of each other, as may be shown by drawing a magnet several times through the mixture, when the iron filings will adhere to the magnet, and the sulphur be left behind.

Exp. 4.—Sand and sugar may be mixed together in any proportions, and the resulting mixtures will possess both the grittiness of the sand and the sweetness of the sugar, each in a modified form; but no action having taken place between them, they may be easily separated by the mechanical act of solution and filtration. The sand will be left behind on the sieve or filter.

Gunpowder affords an exceedingly good instance of the difference between the effects of mechanical mixture and chemical combination. It is made from saltpetre or nitre (nitrate of potassa), charcoal, and sulphur, which are mixed together by mechanical means in the most intimate manner possible; but no chemical action having taken place between them, they still remain separate and distinct. The *nitre* may be washed out by means of water, and by evaporating the water may be again obtained in the solid form. So in like manner the *sulphur* may be washed out by *carbonic disulphide*, and on allowing the disulphide to volatilize, the sulphur may be obtained, while the charcoal remains behind undissolved. If, however, we cause the materials to enter into true chemical combination, all is changed—the mixture is fired by heat, the dormant chemical force is called into being, the three solids disappear and are suddenly converted into an enormous volume of gaseous matter, and new substances are produced, possessing properties totally distinct from those of either the *nitre*, *sulphur*, or *charcoal*.

It is the study of this *chemical force*, and of the laws

which govern its action, which is especially the province of **Chemistry.**

14. Characters of Chemical Action.—Chemical attraction or **affinity** is distinguished from all other kinds of force by several well-marked features.

1. It acts only between particles which are absolutely in contact. It is rarely possible, by mechanical means, to bring the particles in sufficiently close contact for chemical action to commence; recourse is generally had to solution, or fusion, or chemical means.

Exp. 5.—Mix together ½ oz. sodic carbonate and 1 dram tartaric acid, no action takes place, not even if the mixture be ground in a mortar; now place it in a glass and add water, a brisk effervescence ensues, due to the chemical action set up, and the escape of a gas called carbonic anhydride.

Exp. 6.—Mix together iron filings and powdered sulphur, no action takes place; put them in a crucible and apply heat, so as to melt the sulphur, combination ensues, attended with a great manifestation of heat.

Exp. 7.—A mixture of oxygen and hydrogen gas, in the proportion of one volume of oxygen to two volumes of hydrogen, may be made, and left in a soda-water bottle for any length of time, no action will take place between them; but, if an electric spark be passed through, or a light be applied, combination takes place suddenly, and with great violence.

Some solids *appear*, at first sight, able to act chemically on each other, *e.g.*—

Exp. 8.—When *iodine* is sprinkled over *phosphorus*, combination instantly takes place, attended with the evolution of light and heat; but, in this case, both substances are volatile, and the commencement of the action takes place between their vapours.

2. *Chemical action is most strongly exerted between dissimilar substances.** Thus, no action whatever takes place between two pieces of iron, or two pieces of copper, or two pieces of sulphur; but between iron and sulphur, or copper and sulphur, very intense chemical action will take place.

* This circumstance—viz., that chemical action is strongest between *unlike* substances—makes the term "*affinity*," which is commonly employed to express "*chemical action or attraction*," an objectionable one.

As a rule, the greater the difference in the properties of two bodies, the greater is their tendency to mutual chemical action. Chemical union may take place between bodies allied to each other in properties, but such unions are very unstable.

3. *The most striking and characteristic feature of chemical attraction is the entire change of properties with which it is attended,* a change that by no possible reasoning could have been predicted. Under its influence solids are converted into liquids, liquids into solids, and gases into liquids and solids; while all the various changes of colour, taste, and smell, which we meet with on every hand, are entirely due to chemical force or attraction.

EXP. 9.—Triturate or mix in a mortar some freshly crystallized *sulphate of soda* and *carbonate of potash;* the two solids will become converted into a liquid.

EXP. 10.—Take a saturated solution* of *chloride of calcium,* and drop into it a small quantity of *sulphuric acid;* the two clear liquids will become converted into a white opaque solid.

EXP. 11.—Make a strong syrup, by dissolving a small quantity (five or six lumps) of white sugar in a beaker glass, with a little warm water. Place the beaker glass in a soup plate or gas tray, and add gradually strong *sulphuric acid,* and stir; in a few minutes the clear syrup will blacken, begin to effervesce, and rise in the glass, and finally it will become solid (sufficiently so for the stirrer to stand upright), and flow over, filling the soup plate.

The *rationale* of the above experiment is, that sugar is a compound of carbon, oxygen, and hydrogen, the two latter being always present in the proportion in which they would form water. Now sulphuric acid has the property of being very greedy of water whenever it meets with it, and it not only absorbs the water which was used in converting the sugar into syrup, but also that which enters into the composition of the sugar itself, so that the carbon is set free, producing the solid mentioned. The reason of its occupying so large a space is, that, being in a very minute state of subdivision, it is exceedingly porous, and is largely permeated by the liquid and steam produced by the mixture of the sulphuric acid and water.

* Solids do not dissolve in liquids in indefinite proportions; and, when the liquid has dissolved as much as it is capable of doing, the solution is said to be saturated.

If the gases oxygen and hydrogen be mixed in the proportion of one of the former to two of the latter, and exploded in a suitable apparatus, water is produced. The full description of this will be given in Chapter IX. under the heading of *water*.

Exp. 12.—Dip a clean feather into *hydrochloric (muriatic) acid*, and moisten with it the interior of a glass jar, cover immediately with a glass plate, and in like manner moisten the interior of a similar jar with *ammonia;* bring the two jars together mouth to mouth, but with the glass plate between them, they will both appear empty; but if the glass plate be removed, the whole included space will be filled with a dense white vapour, which at last settles on the sides of the jars in the form of a white powder—solid chloride of ammonium.

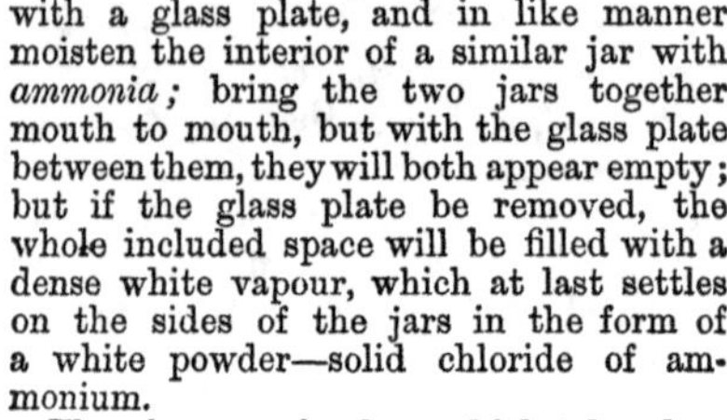

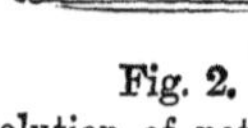

Fig. 2.

The changes of colour which take place under the influence of chemical attraction are so diverse that it is difficult to select instances for illustration, they are so abundant. The arts of dyeing and calico printing are entirely chemical ones.

Exp. 13.—Take four conical test glasses, and in each put as follows:—In the first, a small quantity of *argentic nitrate (nitrate of silver);* in the second, a small quantity of plumbic nitrate *(nitrate of lead);* in the third, some mercuric chloride; and in the fourth, some chlorine water, to which a few drops of a weak solution of starch has been added. (These salts should all be in solution.) The liquids in the four glasses will all be colourless and clear like water; now add to each a few drops of solution of potassic iodide. In the first glass, a pale yellow or straw coloured precipitate will be thrown down; in the second, a dark yellow, almost orange, precipitate; in the third, a brilliant scarlet precipitate; while the liquid in the fourth glass will be turned a beautiful blue.

Many other beautiful and effective experiments on the changes of colour caused by chemical action will be met with in the course of reading this book. The common holiday trick of appearing to pour from a bottle,—previously filled with water,—milk, port, sherry, champagne, blue ink, &c., is done by the glasses being previously prepared with the necessary chemicals to produce the desired effect.

The changes in taste and smell produced by chemical action are very striking, but do not so easily admit of experimental demonstration. Two examples may however be taken.

EXP. 14.—Mix together in a mortar equal parts of *ammonic chloride* and *quick lime*, both of which substances are inodorous. The mixture disengages a gas possessing a most pungent odour.

EXP. 15.—*Chlorine* is a gas, possessing highly irritating, pungent, and poisonous properties. Sodium is a metal of an exceedingly caustic and poisonous nature; and yet if they be caused to unite, they produce a white solid *(common salt, chloride of sodium)*, not only *not* poisonous, but actually necessary to life.

EXP. 16.—If 37 parts of *hydrochloric acid* be mixed with 41 parts of *caustic soda*, both of which substances are intensely poisonous, we shall obtain a neutral solution, consisting of 60 parts of common kitchen salt, dissolved in 18 parts of water.

The only property which chemical action is powerless to alter or even modify is that of *gravity* or *weight*. 16 grs. or oz. or lbs. of *oxygen* will unite with 2 grs., oz., or lbs. of hydrogen, and produce exactly 18 grs., oz., or lbs. of water. So, in the conversion of iron into iron oxide or rust, 7 lbs. of *iron* require exactly 3 lbs. of *oxygen*, and produce 10 lbs. of iron rust. There is *never* a loss of weight in any chemical action. One of the great truths which *Chemistry* teaches is, that, under no circumstances whatever, can there be either creation or destruction of *matter*. In all cases in which the matter *seems* to be destroyed,—as the burning of a candle and other cases of combustion,—the destruction is *apparent* only; the matter under the influence of the chemical force is made to take the form of an invisible gas, and therefore *appears* to be lost. If, however, we take means to collect the products of the combustion, we shall find that so far from losing in weight, they weigh more than the original candle did, the increase in weight being due to the oxygen of the air consumed in the burning. The following simple experiment will show this—

EXP. 17.—A glass tube (Fig. 3)—a common lamp glass does very well—with a cork fitted to the bottom, through which are several holes, in one of which a taper is fixed; the upper part of the lamp glass is connected, by india-rubber tubing, with a

U-tube, in which is placed some caustic soda, the other branch of the U-tube is also connected with a water bottle, or aspirator. The lamp glass, with the candle, the connection, and the U-tube, with the caustic soda, are then carefully weighed, and their weight to the nearest grain accurately noted. The connection with the water bottle being then made, the tap is turned on, and as the water runs out, the air passes up through the lamp glass, &c., to supply its place, and a current or draught of air is thus established through the whole apparatus, the taper is then lit, and it and the cork quickly replaced. When the taper has burnt for a few minutes, the water is turned off, and the taper almost immediately goes out. If, now, the tubes which were weighed at the beginning of the experiment, be again weighed, it will be found that although the candle is considerably less than it was at first, the tubes are perceptibly heavier.

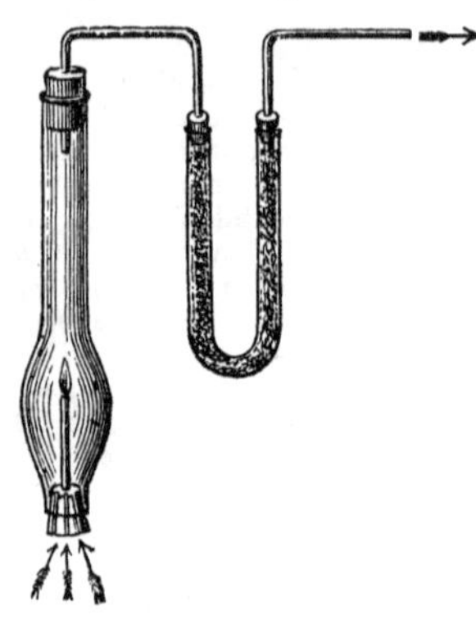

Fig. 3.

The *rationale* of the above experiment is this—the candle is a compound of *hydrogen* and *carbon*, the air is a mixture of two gases, *oxygen* and *nitrogen;* the burning of the candle, like all cases of combustion, is simply a case of chemical combination; so that, in the burning, the *hydrogen* (H) of the candle unites with the oxygen (O) of the air to form water (H_2O) in the condition of steam, and the *carbon* (C) of the candle unites with the *oxygen* of the air to form *carbonic acid* (CO_2). These two invisible gases, *steam* and *carbonic acid*, carried upwards by the draught, are caught and retained in the U-tube by the caustic soda with which they enter into eager combination, and the whole apparatus weighs heavier than at first, by exactly the amount of oxygen which has been abstracted from the air to form the steam and carbonic acid.

Fig. 4.

That *water* and *carbonic acid* are produced when a taper or candle is burnt in the air may be shown by the following experiments :—

Exp. 18.—Hold over the flame of the taper or candle a cold, dry, bright, tumbler or beaker, the inner surface will at once become dimmed by the dew or moisture which will collect on it.

The experiment might be arranged, by keeping the outside of the tumbler or beaker continually cool, so as to collect quite a wine glass of water from the burning candle.

Exp. 19.—Let the taper be burnt in a clean glass bottle with a narrow neck, or with the neck nearly closed by having a piece of narrow glass tubing inserted through the cork. In a few minutes, according to the size of the bottle, the taper will burn dim, and finally go out. If, now, some lime water* be poured in, and the bottle be agitated, the lime water will be turned milky, indicating the presence of carbonic acid—the milkiness being due to chalk which is formed from lime and carbonic acid.

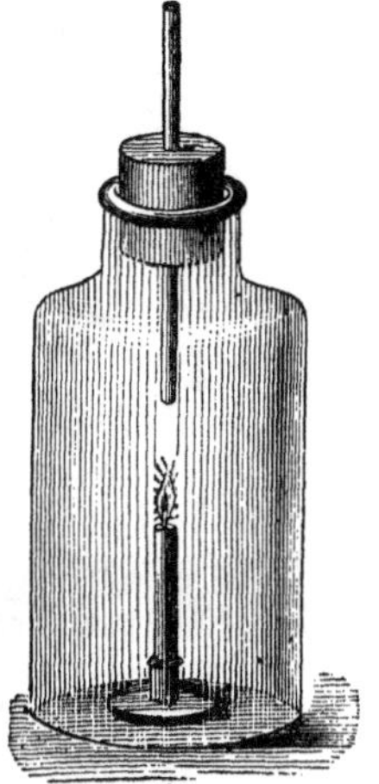

Fig. 5.

4. There are two modes of forming chemical compounds—the simplest, where two substances unite directly together, as when *hydrogen* (H) is burnt in *oxygen* (O) and water is formed, or where an *acid* and an *alkali*, as *hydrochloric acid* (HCl) and ammonia (H_3N), unite together and form a salt. This only takes place when the bodies have a powerful tendency to unite. The other and more common mode of combination occurs when one of the ingredients of a compound is displaced by another substance, and a new compound is the result; for example, if to a solution of *nitrate of lime* (which is clear and transparent as water) *sulphuric acid* be added, the *sulphuric acid* drives out the

* Lime water may be made, by taking a piece of freshly slaked lime, about the size of a walnut, and putting in a wine bottle of distilled water, and shaking well. After standing twenty-four hours the clear liquid is fit for use.

nitric acid and unites with the *lime*, forming *sulphate of lime*, which, being insoluble, falls to the bottom of the glass as a solid white precipitate.

5. Whenever chemical combination takes place as the result of a direct union, heat is always given forth, and the quicker the combination takes place the greater is the heat evolved, in some cases rising so high as to give rise to ignition and combustion—since all solid substances, when heated sufficiently, become luminous.

Exp. 20.—If water be poured on newly burnt lime, the lime swells, breaks into powder, and gives forth intense heat, sometimes sufficiently so to set fire to wood-work with which it may be in contact. Barges and houses have been frequently set fire to from this cause.

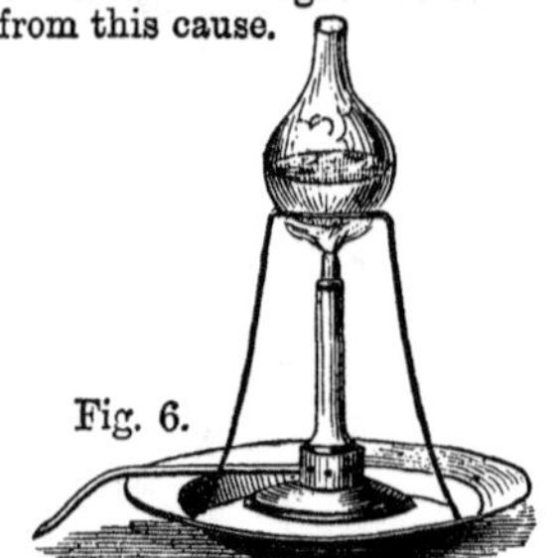

Fig. 6.

Exp. 21.—Place in the bottom of a flask or test-tube about ½ oz. of powdered sulphur, and above that put some copper turnings, previously made hot; now apply heat to the sulphur in the bottom of the flask, which will melt, and finally boil. As soon as the vapour of the burning sulphur reaches the copper, the latter becomes red hot, burns with a lurid red light, melts, and falls to the bottom of the flask, uniting with the sulphur, and forming a black sulphide of copper.

Exp. 22.—If a piece of the metal *potassium*, about the size of a pea, be thrown on the surface of cold water, it will combine with the *oxygen* of the water, setting free the *hydrogen;* and the combination will take place with sufficient intensity to set fire to the *hydrogen*, which burns with a pale violet flame, owing to the presence of a small quantity of the vapour of the potassium.

6. In a mechanical mixture, the proportion of the ingredients may be varied at pleasure, within all conceivable limits, while chemical action takes place only between *certain invariable proportions* of the constituents; thus, if *mercury* (Hg) be heated in *oxygen* gas, red scales will be formed on its surface *(oxide of mercury)*, owing

to its union with *oxygen.* If this red powder be collected and weighed, it will be found that for every 216 grains of the powder the mercury will have lost 200 grains in weight. If, as will sometimes under certain circumstances happen, a black powder be formed instead of a red one, then, on collecting the powder and weighing it, it will be found that the loss of the *mercury* is to the powder formed as 400 grains of mercury is to 416 grains of the black powder. In other words, the metal *mercury* can be made to unite with the gas *oxygen,* by the application of heat in the two following proportions:—

(1.) 400 parts of *mercury* will take up 16 parts of oxygen, and give rise to a black powder known as *sub-oxide of mercury* (Hg_2O).

(2.) 200 parts of *mercury* will unite with 16 parts of *oxygen,* giving rise to a red powder known as *oxide of mercury* (HO).

But no power with which we are acquainted can make mercury and oxygen unite in other than these two proportions.

15. Summary of Chief Characters of Chemical Combination.—The leading characters by which *chemical attraction* is distinguished from all other forces may be thus summed up :—

16. "**Chemical attraction** is a power of extreme energy, which acts only on the minutest particles of matter, and at distances too small to be perceptible. Under its influence the elementary bodies, though comparatively few in number, arrange themselves into the numberless compounds which constitute the different forms of matter in the three great kingdoms of nature; and it is important to observe that the proportions in which they unite are fixed and invariable." That it takes place between different kinds of matter with different but definite degrees of force, and that, as a rule, the greater the difference between the substances the more intense is their chemical action on each other. That it is attended by a total change of properties, both physical and chemical, with the exception of that of gravity or weight. The compound resulting from a chemical union possesses properties totally differing from those of either of its constituents. The reason of gravity or weight remaining unaffected is, that in *chemical action* there is no loss, no destruction of matter, but merely a change of form. And, lastly, that where chemical action is the result, not of substitution, but of direct union of the elements, heat, and not unfrequently light, is manifested; the heat being in all cases a direct measure of the intensity of the chemical action which takes place.

CHAPTER III.

Combining Weights—Laws of Chemical Combination—Outline of Atomic Theory—Combining Volumes—Volume Weights.

17. Combining Weights.—In the last chapter it was stated as one of the peculiar features of chemical attraction that, when it took place, it only occurred between certain fixed, definite, invariable proportions of the constituents. These fixed, definite proportions, ascertained by the comparison of a large number of experiments, and reduced to their simplest and lowest expression, are called the *combining weights* of the substance. As it is necessary to have some standard to which to refer them, chemists are now nearly universally agreed to call the lowest proportion of *hydrogen* 1, and to refer the weights of all others to that scale. Thus, the *combining weight* of *hydrogen* being 1; that of *oxygen* becomes 16; of *nitrogen*, 14; of *carbon*, 12; of *sulphur* 32; of *mercury*, 200; of *zinc*, 65; of *iron*, 56, and so on. A complete list of the *combining weights* of all the *elements* is given on pages 13 and 14, with the table of the *elements*.

18. Laws of Chemical Combination.—The relative

proportions in which the elements unite are regulated by fixed laws. These laws (four in number) form the basis of all chemical science, and regulate the mode of combination of every known chemical substance, simple and compound. They are called the laws of chemical combination.

19. Law 1.—This is called the **law of constant composition,** or **law of definite proportions.** It may be enunciated as follows:—"*The same chemical compound always contains the same elements, united together in the same proportions.*"

Thus, 100 parts of water always contain $88\cdot\dot{8}$ ($88\frac{8}{9}$) parts of *oxygen*, and $11\cdot\dot{1}$ ($11\frac{1}{9}$) parts of *hydrogen*. The weight of the oxygen is always eight times that of the hydrogen. It does not matter from what source the water is taken, whether it be from the melting of snow on the tops of the highest mountains, from rain clouds, from dew, or from that obtained by chemical action, its composition is invariable. It is impossible for us to conceive water with the same properties as ordinary water, but having a different proportion of oxygen and hydrogen from that above given. So, also, a piece of flint or rock crystal, no matter where it may come from, will be always found on analysis to yield in 100 parts 46·6 parts of silicon, and 53·4 parts of oxygen. In fact, experiment shows that all true chemical compounds, which have been submitted to analysis, have a fixed definite composition.

It is the generality and universality of this law that give to analysis its practical value, since the results are always uniform and certain.

The converse of this law, however, viz.: "That the same chemical elements united together in the same proportions will always produce the same chemical compound"—does not by any means hold good.

20. Isomerism.—A great many substances have been discovered amongst organic bodies composed of the same elements in the same relative proportions, and yet exhibiting physical and chemical properties perfectly distinct one from another—*e. g.*, oil of lemons, oil of turpentine, oil of rosemary, and many others all contain the same elements, united together in the same proportion; so the crystallized portion of *essence of roses* and common *coal gas* are chemically identical. To all such bodies the term **Isomeric** (from ἰσος, *equal*, and μερος, *part*) is applied.

21. Law 2, called the **law of multiple proportions,** may be thus enunciated—"*When two bodies are capable of uniting together in more than one proportion, these proportions will bear a simple ratio to each other.*"

When simple elementary bodies unite in more than one proportion, the compounds so obtained will differ altogether in chemical properties, but there will be a regularity in the plan on which they are formed. Thus, if A and B be the elements in question, if A be constant, B will be found to exist in a series of multiples, such as—

A + B, A + 2 B, A + 3 B, A + 4 B, &c., or
A + B, A + 3 B, A + 5 B, &c., or
2 A + B, 2 A + 2 B, 2 A + 3 B, &c., or
2 A + 3 B, 2 A + 5 B, 2 A + 7 B, &c.,

or in some similar series. This is very fully exemplified in a large number of well known compounds—*e. g.*, those of oxygen (O) and hydrogen (H), of which there are two, viz.:

Water consists of 2 hydrogen and 1 oxygen, H_2.
Hydroxyl ,, 2 ,, 2 ,, H_2O_2 or 2 HO.

Mercury and *chlorine* form two distinct compounds, viz.:

Calomel, consisting of 200 parts of *mercury* and 35·5 parts of *chlorine*;
Corrosive sublimate, consisting of 200 parts of *mercury* and 71 parts of *chlorine*;

or one combining weight of *mercury* to one combining weight of *chlorine*, and one combining weight of *mercury* to two combining weights of *chlorine*. The most instructive example of this rule is afforded by the series of compounds formed by *nitrogen* and *oxygen*. They are five in number, and the oxygen proceeds with marked regularity. The analysis of 100 parts of each gives the following results :—

	Nitrogen.	*Oxygen.*	*Nitrogen. Oxygen.*	Symbol.
Nitrous Oxide,......	63·64	36·36	:: 28 : 16	N_2O
Nitric Oxide,........	46·67	53·33	:: 28 : 32	N_2O_2
Nitrous Anhydride,	36·85	63·15	:: 28 : 48	N_2O_3
Nitric Peroxide,....	30·44	69·56	:: 28 : 64	N_2O_4
Nitric Anhydride, .	25·93	74·07	:: 28 : 80	N_2O_5

Thus it will be seen that, while the proportion of *nitrogen* remains constant, being 28, or twice its combining weight 14, that of *oxygen* increases regularly, being 1, 2, 3, 4, and 5 times its combining weight 16.

In most cases, the proportion is not always so simple, two parts of one element uniting with 3, 5, or 7 parts of another.

This important law was first clearly established by Dalton, and has been made by him the foundation of his **atomic theory,** which we have yet to consider.

22. Law 3.—The third law of chemical combination is usually known as **the law of equivalent proportions.** It is sometimes called **the law of combining proportions of elements.** It may be thus stated, "*Each elementary substance, in combining with other elements, or in displacing others from their combinations, does so in a fixed proportion, which may be expressed numerically.*"

This law may be also expressed in another way, thus —"If a body, *A*, unite with certain proportions by weight of other bodies, *B*, *C*, and *D*, those proportions, or multiples, or sub-multiples of them, will represent also the proportions in which *B*, *C*, and *D* will unite among themselves, or with any other bodies, *E*, *F*, *G*, *H*, &c."

It is the full consideration of this law which has enabled us to determine the "*combining or atomic weights*" given in the first chapter.

We are here met with a difficulty in distinguishing between "*combining or atomic weight*" and *equivalent*,—a distinction, which we shall not fully be able to understand until we have discussed the doctrine of **equivalence** or **atomicity.**

The meaning of this *equivalence* will be seen on comparing the series of chlorides, as follows :—

Hydrochloric acid, or *Chloride of hydrogen*,	is composed of	35·5 parts chlorine,	and 1 part	*hydrogen.*
Chloride of zinc, ...	,,	35·5	,,	32·5 ,, *zinc.*
Chloride of silver, ...	,,	35·5	,,	108 ,, *silver.*
Chloride of copper, ...	,,	35·5	,,	31·75,, *copper.*
Chloride of mercury, ..	,,	35·5	,,	100 ,, *mercury.*

So that, in power of combination with the same quantity, 35·5 parts of *chlorine*, 1 of *hydrogen*, 32·5 of *zinc*, 108 of *silver*, 31·75 of *copper*, and 100 of *mercury*, seem to be equal, and might be regarded as *equivalents.*

But *mercury* and *copper* both form other chlorides, containing twice the amount of metal above given, viz., 200 parts of *mercury*, or 63·5 parts of copper, will each combine with 35·5 parts of *chlorine.* These two metals then would seem to have each two equivalents. Which is the true one, and which is to be taken as the *combining* or *atomic weight?* In all such ambiguous cases the true combining weight is found by a comparison of the combinations of the elements in question with other elements, and notably with *oxygen;* in the cases in point *mercury* is never found combining with *oxygen* in a lower ratio than 200 to 16, and *copper* in a lower ratio than 63·5 to 16. The numbers then, 200 and 63·5, are taken as the combining weights of mercury and copper respectively.

23. Law 4.—The fourth law of chemical combination

may be stated thus—"*The combining weight of a chemical compound is the sum of the combining weights of its components.*" Thus—

Hydrochloric acid (which is made up of one part of *hydrogen*, and one part of *chlorine*), has for its combining weight 36·5, which is obtained as follows:—

1 part of *hydrogen*, ...	=	1
1 ,, *chlorine*, ...	=	35·5
Hydrochloric acid, ...	=	36·5

So *sulphuric acid* (which is made up of 2 parts of *hydrogen*, 1 part of *sulphur*, and 4 parts of *oxygen*,) has for its combining weight 98, obtained as follows:—

2 parts of *hydrogen*, ...	=	2
1 ,, *sulphur*, ...	=	32
4 ,, *oxygen*, ...	=	64
Sulphuric acid, ...	=	98

And in like manner for all other compounds. This law, which necessarily follows from the other laws, and from the fact that in chemical combination no loss of weight takes place, is nevertheless very important in a practical point of view, since by it we are able to regulate exactly the amount of any chemical compound we must employ in order to produce a given chemical effect, without loss or waste of material.

24. Atomic Theory.—The *atomic theory*, for which we are indebted to Dalton, presupposes matter to be composed of ultimate undivisible particles or *atoms* (from α, *not;* τεμνω, *I cut*), which unite together in various proportions; that these atoms are in the same element, exactly similar in size, weight, and every other property; that the atoms of any one element differ from those of all other elements in weight and chemical properties; and when union takes place, it must of necessity take place between atom and atom, or between a definite number of atoms of both elements. It would follow, then, that granting the existence of such atoms, and that the atoms of any one element must be equal in weight, while the atoms of different elements would differ in weight, chemical combination, if it did take place, would do so in certain well marked definite proportions by weight, viz., the relative weights of the different atoms, or in some multiples of those weights. All observation and experiment show us that chemical combination does so take place.

As we cannot explain the facts of chemical combination by any other hypothesis, these facts become a strong "*a priori*" proof of the truth of the *atomic theory*. Still, it can never be more than theory, since it would be impossible that we should ever succeed in isolating an atom, and thus obtain direct proof; while, mentally, we cannot conceive of any particle, however small, but that we can also conceive of its half, or any fraction of it; and mathematically, it is possible to demonstrate that space, and therefore matter, which occupies space, is capable of infinite subdivision.

Thus, let A, B, C, D be lines drawn parallel to each other; draw the oblique line FG, and from F on the indefinite right line CD

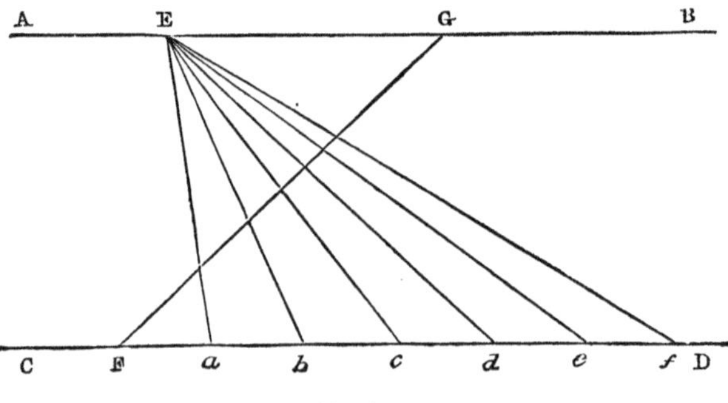

Fig. 7.

take any number of equal parts, as F, *a*, *b*, *c*, *d*, &c. From any point E in the straight line AB draw lines connecting this point with *a*, *b*, *c*, *d*, &c., cutting the oblique line FG; then, as the number of points, *a*, *b*, *c*, *d*, on the line CD may be infinite, it follows that the line FG may be infinitely divided, by lines connecting such points to E.

Matter has been *actually* divided to an extent that is positively inconceivable, although not inappreciable nor immeasurable, and yet we are not able to assert that we have arrived at the ultimate limit; we can only say that our physical means of division and measurement are for the present exhausted, *e.g.*—

"In the ordinary process of making gold leaf, for example, the gold is hammered out so thin that 280,000 leaves would be required to make up the thickness of an inch, and a single grain of gold is hammered out until it covers a square seven inches in the side. Each square inch of this may be cut into 100 strips, and each strip into 100 pieces, each of which is distinctly visible to the unaided eye. A single grain of gold may thus, by mechanical means, be subdivided into $49 \times 100 \times 100 = 490{,}000$ visible pieces. But this is not all; if attached to a piece of glass this gold leaf may be subdivided still further; 10,000 parallel lines may be ruled in the space of one single inch, so that a square inch of gold leaf, weighing $\frac{1}{49}$ of a grain, may be cut into 10,000 times 10,000, or 100,000,000 pieces, or an entire grain into 4,900,000,000 fragments, each of which is visible by means of the microscope. Yet we are quite sure that we have not even approached the possible limits of subdivision, because, in coating silver wire, the covering of gold is far thinner than the gold leaf originally attached to it, since in drawing down the gilt wire the gold continues to become thinner and thinner each time, in proportion as the silver wire itself is reduced in thickness."

"When a substance is dissolved in any liquid, the subdivision is carried still further, and the particles are rendered so minute as to escape our eyesight, even when aided by the most powerful magnifiers."—Miller's *Chemical Physics*, page 5.

It may be, and probably always will be, impossible to actually demonstrate the physical existence of an atom; but as a celebrated modern chemist has remarked, "*that whether matter be atomic or not, thus much is certain, that granting it to be atomic, it would appear as it now does.*"

The late Professor Faraday sums it up in these words—"Seeing that all marked cases of chemical combination can be demonstrated always to take place in definite proportion, and that, by inference, a similar proportionality may be supposed to extend to less marked cases,—seeing that these definite proportions of bodies entering into combination are mutually proportional amongst themselves, it follows that such definite immutability, such proportionality, should most rationally be considered as indicating a ponderable ratio between combining elements; and that the ratio never changing would seem to be indicative of elementary ponderable molecules of determinate relative weight, unchanging, indivisible, qualities which will be recognised as fulfilling the definition of an atom."

Granting, then, (1.) The existence of atoms; (2.) That the atoms of the same element are absolutely identical in size, weight, and all other respects; (3.) That the atoms of one element differ from those of another element in weight and chemical properties; and (4.) That whenever combination takes place between two elements, it occurs between them atom to atom—Miller draws these three conclusions :—

"*First*, That the proportion in which combination occurs must, when the same compound is formed, always be *definite*, since that proportion is determined by the relative weights of the atoms of the combining elements, and the atom cannot be subdivided."

"*Second*, That when the same elements unite in several proportions, these proportions must vary according to the terms of a *simple series of multiples*, since each atom of one element must unite with the other element in the ratio of one, of two, or of three atoms, or in some other ratio equally simple, inasmuch as the atom does not admit of subdivision."

"*Third*, That combination must occur also in *equivalent* proportion, since the equivalent amounts of each element must be in the proportion either of the weights of their atoms, or of a simple multiple of those weights."

The chemist then uses the term "*atom*," much as he uses that of "*element*," not as expressing an absolute fact, but as a convenient term to express what is observed to be the case according to our present knowledge.

25. Combining Volumes.—When bodies are capable of assuming the form of gas or vapour, a very simple relation exists between the volumes or bulks of those gases which combine together and the bulk of the gaseous compound formed by their union. The law is known as **Gay Lussac's law of volumes.** It is found that gases or vapours unite together by volume either in the proportion of equal bulks or in that of one volume to two volumes, or one to three volumes, or in some very rare cases in that of two to three. Thus equal volumes of *hydrogen* and *chlorine* unite together to form *hydrochloric acid;* two volumes of *hydrogen* unite with one volume of *oxygen* to form *water;* and three volumes of *hydrogen* unite with one volume of *nitrogen* to form *ammonia.*

26. Volume Weights.—This arises from the fact, that if equal quantities or volumes of the elements in the state of gas or vapour be taken, their weights will be found to be in the ratio of their atomic weights.* Thus—

44·4	*cubic inches*	of	*hydrogen*	weigh	1	grain.
44·4	,,	,,	*oxygen*	,,	16	,,
44·4	,,	,,	*nitrogen*	,,	14	,,
44·4	,,	,,	*chlorine*	,,	35·5	,,

and so on.

Or putting it in French measure (see next chapter)—

11·19	litres of	*hydrogen*	weigh	1	gramme.
11·19	,,	*oxygen*	,,	16	,,
11·19	,,	*nitrogen*	,,	14	,,
11·19	,,	*chlorine*	,,	35·5	,,

and so on.

After union, the volume of the resulting compound, though frequently less than the combined volume of its constituents, bears, nevertheless, a simple relation to it. When gases combine together in equal volumes they generally undergo no change of volume; sometimes, however, the two volumes become condensed to one; in other cases the three volumes become condensed to two, as in the case of water, or even to one. In no case, however, do the combined gases occupy a larger space than they did when separate.

In estimating the relation between the volume and weight of all gases and vapours, both simple and compound, reference must always be made to the temperature and pressure to which they are subjected.

Heat has the property of expanding all things, but gases and vapours expand very largely for any increase of

* An exception to this law obtains in the case of the vapours of *phosphorus, arsenic, mercury, zinc,* and some other less important elements.

temperature. It has also been found, by accurate experiments, that *gases expand regularly for every increase of temperature,* the rate of expansion being $\frac{1}{273}$ of their volume at zero or freezing point, for every increase of 1° centigrade. Thus—

273	volumes or measures of air or gas at 0° C. become				
274	,,	,,	,,	,,	1° C.
275	,,	,,	,,	,,	2° C.
273 + x	,,	,,	,,	,,	x° C., and so on.

The fraction $\frac{1}{273}$ is represented by the decimal fraction 0·003665, and so—

1	volume of air or gas at 0° C. becomes			
1·003665	,,	,,	,,	1° C.
1·007330	,,	,,	,,	2° C.
1·010995	,,	,,	,,	3° C.,

and so on, the pressure in all cases remaining the same.

But the volume which a given quantity of gas occupies at any temperature depends also on the pressure. If the pressure under which the gas exists be removed or lessened, the gas will immediately *increase* in size or volume; and if the pressure be increased, the volume of the gas will be diminished regularly. The law according to which this takes place was discovered by Boyle in England, and by Mariotte on the Continent, independently of each other, and known as the **Law of Boyle and Mariotte.** It may be thus enunciated—"*The temperature remaining the same, the volume which a gas occupies varies* INVERSELY *as the pressure to which it is subjected;* or, *the density of a gas is proportionate to its pressure.*" Thus—

1 volume	or measure of air or gas under 1 pressure will become			
2 vols.	,, measures	,,	$\frac{1}{2}$	,,
3 vols.	,, measures	,,	$\frac{1}{3}$	,,

Again—

1 vol.	or measure of air or gas under 1 pressure will become			
$\frac{1}{2}$ vol.	,, measure	,,	2	,,
$\frac{1}{3}$ vol.	,, measure	,,	3	,,

and so on, provided the *temperature* remains the same.

If the *temperature* and the pressure both vary, the volume of the gas will vary, directly as the temperature, and inversely as the pressure.

The standard temperature and pressure at which the volumes of gases and vapours are recorded, are 62° F., and barometer at 30 inches; or, in French measure, 0° C., and barometer 760 m.m. (= 29·9 English inches).

If the volume of a given quantity of a gas, at other than the normal temperature and pressure, be required, the corrections for each must be made separately, as in the following examples:—

1. 100 cubic inches of a gas are taken at 15° C. What volume will they occupy if the temperature be raised to 20° C.?

Now, since gases expand $\frac{1}{273}$ of their volume at 0° C. for each degree centigrade,

$$\therefore 273 \text{ vols. at } 0^\circ \text{ C.} = 288 = (273+15) \text{ vols. at } 15^\circ \text{ C.}$$
$$\text{and } 273 \text{ vols. at } 0^\circ \text{ C.} = 293 = (273+20) \text{ vols. at } 20^\circ \text{ C.}$$

vol. at 15° C.		vol. at 15° C.		vol. at 0° C.		vol. at 0° C.
∴ 288	:	100	::	273	:	x

Here x = 94·792 cub. in.

i.e., 100 cubic inches of gas at 15° C. = 94·792 cub. in. at 0° C.

Again—

vol. at 0° C.	vol. at 0° C.	vol. at 20° C.	vol. at 20° C.
vol. at 10° C.	vol. at 0° C.	vol. at 20° C.	vol. at 10° C.

And x = 101·7364 cub. in. Answer.

i.e., You first reduce the given volume of the gas at the given temperature to its corresponding volume at 0° C., and then by means of the fraction $\frac{1}{273}$, you can find its volume at the required temperature.

2. 100 cubic inches of a gas are taken when the barometer stands at 31 inches, what volume will it occupy if the barometer sink to 28 inches?

The observed volume is taken at 31 inches, while the required volume is to be at 28 inches, and as gases expand inversely as the pressure, the volume at 28 inches bar. $= \frac{31}{28}$ of vol. at 31 inches bar.

$$\frac{31}{28} \text{ of } 100 = \frac{31}{7} \text{ of } 25 = \frac{775}{7} = 110 \cdot \dot{7}1428\dot{5} \text{ cub. inches.}$$

This law of variation for temperature and pressure holds good for all gases, and for atmospheric air. The modification necessary for vapours, and for condensible gases when near their point of condensation, will be explained in the *Advanced Series.*

CHAPTER IV.

French and English Systems of Weights and Measures—Conversion of English into French Weights and Measures—The Crith and its Uses.

27. THE accurate use of the balance, or in other words, the determination of the size and weight in which substances enter into combination, cannot be over-estimated in chemistry.

By **weight** we understand the force with which **gravity** acts, and which at the same place is always constant.

The selection of a standard of comparison for this force is purely arbitrary. In England we make use of a certain weight called a pound Avoirdupois; this we subdivide into 7,000 parts, which we call *grains*, and we have further agreed to call 5,760 of these grains a pound Troy (the weight used for gold, silver, or precious stones). In chemistry it is customary to use Avoirdupois weight.

The system of measures is connected with that of weight by the definition of a gallon as a measure which shall contain 10 lbs. of distilled water, at a temperature of 60° F.

(15·5 C.), and the barometer standing at 30 inches. A gallon of distilled water thus weighs 70,000 grains.

These measures are, on the other hand, connected with those of length, by the determination that a gallon contains 277·276 cubic inches; consequently a cubic inch of water at 60° F., barometer at 30 inches, weighs 252·4597 grains, or, in round numbers, 252·46 grains, or very nearly $252\frac{1}{2}$ grains.

The French use a decimal system of weights and measures, both of capacity and length. They adopt as the unit of their system the metre (frequently written and pronounced meter in English works); hence this system is generally known as the metric system.

The *metre* (equal to 39·37079 inches, a little more than an English yard), was assumed to be the $\frac{1}{10,000,000}$th part of the earth's quadrant (from the equator to the poles, as determined by French geometricians),* and a bar of metal of this length was carefully prepared and deposited at Paris as a standard metre for future reference.

The divisions of the *metre* are distinguished by the prefixes *deci-*, *centi-*, and *milli-*; so that a *decimetre* is $\frac{1}{10}$ of a *metre*, a *centimetre* $\frac{1}{100}$, and a *millimetre* $\frac{1}{1000}$, of a *metre*.

On the other hand, the multiples of the *metre* are distinguished by the prefixes *deca-*, *hecto-*, and *kilo-*, so that—

10 *metres*	equal to	1 decametre.
100 ,,	,,	1 hectometre.
1,000 ,,	,,	1 kilometre.

The measures of capacity are connected with those of length, by taking as the unit the contents of a cube whose side is equal to *1 decimetre* (3·937 inches). This quantity is called a *litre* (equal to 1·7637 English pints), and has

* Later and more accurate observations have demonstrated that this measurement of the earth's quadrant is in error to the extent of about $1\frac{1}{2}$ miles, so that the metre is very nearly, but not exactly, $\frac{1}{40,000,000}$th part of the earth's circumference. This, however, does not interfere with its value as the unit of the metric system.

for its divisions and multiples the same prefixes as the *metre*, so that—

$\frac{1}{1000}$ *litre*	equal to	1 *millilitre* or	*cubic centimetre*, c.c.
$\frac{1}{100}$,,	,,	1 *centilitre* or 10	,, c.c.
$\frac{1}{10}$,,	,,	1 *decilitre* or 100	,, d.c.
10 *litres*	,,	1 *decalitre.*	
100 ,,	,,	1 *hectolitre.*	
1,000 ,,	,,	1 *kilolitre.*	
10,000 ,,	,,	1 *myriolitre.*	

And, lastly, the system of weights is connected with both those of length and capacity, by assuming as the unit the weight of 1 *cubic centimetre* (*c.c.*) of distilled water, at a temperature of 4° C. (39·2 F.) This is called a *gramme* (equal to 15·432 English grains), and is divided into tenths or *decigrammes,* hundredths or *centigrammes,* and thousandths or *milligrammes;* while, in the same way as with the *metre* and *litre,* we have *decagrammes, hectogrammes, kilogrammes,* and *myriagrammes,* for 10, 100, 1,000, and 10,000 *grammes* respectively.

The *litre,* as it contains 1,000 *cubic centimetres* of distilled water at 4° C., is thus equal in weight to one *kilogramme.*

The kilogramme is the commercial measure of weight in use in France.

The *decimal* or *metric system* is found in many respects to be so convenient, that it is universally adopted in all scientific researches abroad, and is daily gaining ground among scientific men in this country. In this little work we shall give all weights and measures in the metric system, but shall add in brackets the nearest approximation in English value.*

28. Conversion of French System into English.—The following table will also materially assist in the conversion of French weights and measures into English:—

* Square measure is so little used in chemistry, that it has been omitted from this work.

	French.	English.		
Weight.	Centigramme, ...	·15432 grains troy,	about $\frac{1}{6}$ grain.	
	Decigramme, ...	1·5432	„	„ $1\frac{1}{2}$ „
	Gramme, ...	15·432	„	„ $15\frac{1}{2}$ „
	Decagramme, ...	154·32	„	nearly $\frac{1}{3}$ oz.
	Hectogramme, ...	1543·2	„	„ $3\frac{1}{4}$ „
	Kilogramme, ...	15432	„	„ $2\frac{1}{4}$ lbs.
Lineal Measure.	Millimetre, ...	·03937 inches,	about $\frac{1}{25}$ inch.	
	Centimetre, ...	·3937	„	„ $\frac{2}{5}$ „
	Decimetre, ...	3·937	„	„ $\frac{1}{3}$ foot.
	Metre, ...	39·37	„	„ $3\frac{1}{4}$ feet.
	Decametre, ...	393·7	„	nearly 11 yards.
	Hectometre, ...	3937·07	„	„ 110 „
	Kilometre, ...	39370·7	„	about $\frac{5}{8}$ mile.
	Myriametre, ...	393707·9	„	„ $6\frac{1}{4}$ miles.
	Millilitre, or Cubic centimetre,	·061 cubic inches,	about $\frac{1}{16}$ cub. in.	
	Litre, ...	61·027	„	„ $1\frac{3}{4}$ pints.
	Decalitre, ...	610·27	„	„ $2\frac{1}{5}$ gallons.
	Hectolitre, ...	6102·7	„	„ 22 „

1 inch = about $2\frac{1}{2}$ centimetres.	1 yard = about $\frac{9}{10}$ metre.
1 foot = „ 3 decimetres.	1 mile = „ $1\frac{3}{5}$ kilometre.
1 cubic inch = about $16\frac{1}{3}$ c.c.	1 cubic foot = about $28\frac{1}{3}$ litres.

1 gallon = about $4\frac{1}{2}$ litres.

1 grain, ...	= about $\frac{1}{15}$ gramme.
1 oz. troy, ...	= „ $31\frac{1}{10}$ „
1 oz. avoirdupois, ...	= „ $28\frac{2}{5}$ „
1 lb. „	= „ $\frac{9}{20}$ kilogramme.
1 cwt. „	= „ $50\frac{4}{5}$ „

29. The Crith and its Uses.—The *crith* (from κριθη, a barleycorn), used figuratively to denote a small weight, is the term proposed by Dr. Hofmann, and since universally admitted, to indicate the absolute weight of 1 litre or cubic decimetre of hydrogen gas at the normal temperature and pressure, viz., temp. 0° C., and pressure, barometer, 760 millimetres of mercury.

Of this hydrogen unit Dr. Hofmann says, "The actual

weight of this cube of hydrogen, at the standard temperature and pressure mentioned, is ·0896 gramme; a figure which I earnestly beg you to inscribe, as with a sharp graving tool, upon your memory. There is probably no figure in chemical science more important than this one to be borne in mind, and to be kept ever in readiness for use in calculation at a moment's notice. For this litre weight of hydrogen = ·0896 gramme (I purposely repeat it) is the standard multiple or co-efficient, by means of which the weight of one litre of any other gas, simple or compound, is computed. Again, therefore, I say, do not let slip this figure, ·0896 gramme."

If we call the weight of 1 litre of *hydrogen* 1 *crith,* as the weight of equal volumes of gases will be in the proportion of their atomic weights, the weight of 1 litre of oxygen = 16 criths, of 1 litre of nitrogen = 14 criths, and of 1 litre of chlorine gas = 35·5 criths, and so on; while their absolute weight in grammes will be obtained as follows:—

1 litre of *oxygen* = 16 criths = 16 × ·0896 grammes = 1·4336 grammes.
1 „ *nitrogen* = 14 „ = 14 × ·0896 „ = 1·2544 „
1 „ *chlorine* = 35·5 „ = 35·5 × ·0896 „ = 3·1808 „

So, in the case of compound gases, 1 vol. of *hydrogen* unites with 1 vol. of *chlorine* to form 2 vols. of *hydrochloric acid* (HCl) = 36·5 by weight. Hence 2 vols. of HCl weighing 36·5, 1 volume must weigh $\frac{36\cdot5}{2}$ = 18·25; the weight therefore of 1 litre of hydrochloric acid gas, at the normal temperature and pressure, equal to 18·25 *criths* = 18·25 × ·0896 *grammes* = 1·6351 *grammes.*

So 2 vols. of *hydrogen* unite with 1 vol. of *oxygen* to form 2 vols. of *water vapour,* which weigh 18 units, therefore 1 vol. of water vapour must weigh $\frac{18}{2}$ or 9 units, and consequently 1 litre of water vapour = 9 *criths* = 9 × ·0896 *grammes* = ·8064 *grammes.*

In the same way, 2 vols. of *hydrogen* and 1 of *sulphur* unite together and form 2 volumes of *hydrogen sul-*

phide; now H_2S measures 2 volumes, and weighs 2 + 32 = 34; therefore 1 volume of H_2S = 17 units, and consequently

1 litre *hydrogen sulphide* = 17 criths = 17 × ·0896 grammes
= 1·5232 grammes.

Lastly, 3 vols. of hydrogen (= 3 criths) unite with 1 vol. of nitrogen (= 14 criths) to form 2 vols. of ammonia (H_3N), (= 17 criths); 1 vol. of H_3N = $\frac{17}{2}$ = 8·5 criths, and therefore—

1 litre of ammonia = 8·5 criths = 8·5 × ·0896 grammes
= ·7616 grammes.

Thus, the actual weight of a litre of any gas, simple or compound, at the normal standard of temperature and pressure, may be obtained by multiplying its atomic weight by ·0896, the standard weight of a litre of *hydrogen*.

CHAPTER V.

Principles of Chemical Nomenclature—Classification of Elements into Positive and Negative—Symbolic Notation—Chemical Formulæ—Chemical Equations.

30. The principle on which the system of **chemical nomenclature** is founded is, that in the case of the elements, the name shall give some idea of the nature, properties, and affinities of the substance; and in that of compounds, that it shall indicate the composition and constitution of the body to which it is applied.

In the case of the elements, with some few exceptions, this object has been attained; but in that of compounds, it is difficult in a science like chemistry, where new discoveries are continually altering our views of the constitution of bodies already existing, or bringing fresh ones to our knowledge, to devise a system of nomenclature which shall at the same time be sufficiently precise to express all existing views, and sufficiently elastic to embrace all new ones. On this account chemical nomenclature may be regarded as in a somewhat unsettled and unsatisfactory condition. In naming many salts, no less than four different usages prevail. Thus the salt, whose composition is expressed by the formula K_2SO_4, is known by the four following names—

Sulphate of potash.	Sulphate of potassium.
Dipotassic sulphate.	Potassium sulphate.

We shall perhaps better be able to explain the most general method on which the systematic names are constructed, if we first draw the distinction which is observed between the elements with respect to their electrical properties, or their division into *basylous* or *electro-positive*, and *chlorous* or *electro-negative*, elements. Chemical compounds are freely decomposed by electricity; when so decomposed, those elements which appear at the positive pole are called electro-negative, while those which appear at the negative pole are called electro-positive. The terms *basylous* and *chlorous* must be left without explanation for the present.

It is somewhat difficult to arrange the elements in an electrical series, seeing no two observers exactly agree.

Dr. Frankland gives the following eight elements as being *negative* or *chlorous* towards the remaining fifty-six elements, which are always more or less *basylous*:—

Fluorine.	Oxygen.
Chlorine.	Sulphur.
Bromine.	Selenium
Iodine.	Tellurium.

It must be borne in mind that the difference between the two classes, the electro-positive and electro-negative, is one of degree only. *Mercury*, for instance, is negative to *sodium*, and positive to *iodine*. The elements may be arranged in such a series that any one in combination is electro-positive to any following, but electro-negative to all preceding ones. The following, table, taken from Ferguson's *Electricity*, page 122, agrees in the main with the order given by Berzelius.

⟶

\+ Potassium. Sodium. Magnesium. Zinc. Iron. Aluminium. Lead. Tin. Bismuth. Copper. Silver. Mercury. Platinum. Gold. Hydrogen. Antimony. Carbon. Phosphorus. Iodine. Chlorine. Nitrogen. Sulphur. Oxygen. —

In this table oxygen is put as the most negative, but later researches have shown it to be almost certain that chlorine and its allied elements bromine, iodine, and fluorine, should stand at the negative end of the series.

31. Nomenclature of Compounds.—The simplest possible chemical compound is one formed by the union of two elements, and called a *binary compound.* The name of the positive element is placed first, with the adjective termination *ic;* and that of the negative element last, with the termination *ide;* thus—

Mercury	and	chlorine	produce	mercuric chloride.
Silver	,,	bromine	,,	argentic bromide.
Potassium	,,	iodine	,,	potassic iodide.
Iron	,,	sulphur	,,	ferric sulphide.
Sodium	,,	oxygen	,,	sodic oxide.

And so on. The same elements sometimes form two distinct compounds, as, for instance, *iron* unites with two different proportions of *chlorine.* To distinguish these, the name, "*Ferrum,*" of the iron (the positive element) is made to terminate in "*ous,*" and "*ic,*" respectively, viz., "*ous*" for the compound, which contains the least quantity of *chlorine* (the negative element), and "*ic*" for that which contains the larger quantity of chlorine; thus—

Ferrous chloride, $FeCl_2$. *Ferric Chloride*, Fe_2Cl_6.*

In cases where more than two compounds are formed

* For explanation of these and succeeding symbols, see page 13.

by the same elements (which is very rare), they are distinguished by the prefixes *hypo-* (beneath), and *per-* (hyper, above).

Many of the binary compounds formed by *oxygen* have the property, when added to water, of acquiring *acid* characters.

The term **acid** was originally applied to all substances which were soluble in water, had a distinctly acid or sour taste, and possessed the property of turning a vegetable blue colour red. Tincture of litmus, which is of a blue colour, is exceedingly sensitive to the action of an acid; and paper stained with this tincture (litmus paper) is the test which the chemist applies to detect the presence of an acid.

All bodies which come under the above definition are still regarded as acids, but the term "*acid*" has now received a much more extended signification.

32. Frankland's Definition of an Acid.—It may be now defined to be a compound containing one or more atoms of hydrogen, which are capable of being displaced by a metal, either partially or entirely.

Thus, *hydrogen* and *chlorine* unite together in equal volumes to form *hydrochloric acid;* if, now, the metal *zinc* be added to this, the acid is decomposed, *zincic chloride* or *chloride of zinc* is formed, and the *hydrogen* is set free. The change is represented by symbols, as follows—

$$2\,HCl + Zn = ZnCl_2 + H_2.$$

Or in the case of acids containing *oxygen*, if the metal be presented in the form of a hydrate, the metal combines with the acid, and water is set at liberty; thus, if sodic hydrate be mixed with nitric acid, sodic nitrate is formed, and water set at liberty. The change may be thus symbolically represented—

HNO_3	+	$NaHO$	=	$NaNO_3$	+	H_2O.
Nitric acid.		Sodic hydrate.		Sodic Nitrate.		Water.

All acids which contain oxygen* have their names formed by adding the termination "*ic*" to the name of the element which is combined with the oxygen, or else to an abbreviation of the name; thus, *sulphur* and *oxygen* form *sulphuric acid; nitrogen* and *oxygen, nitric acid; phosphorus* and *oxygen, phosphoric acid; carbon* and *oxygen, carbonic acid,* and so on. In the case of the element forming two acids with different proportions of oxygen, the one with the lowest proportion of oxygen is distinguished by the termination "*ous,*" while that with the higher proportion receives the termination "*ic.*" Thus, we have *sulphurous acid,* and *sulphuric acid, nitrous* and *nitric acids, phosphorous* and *phosphoric acids,* and so on. When the element forms more than two acids with oxygen, they are distinguished by the prefixes *hypo-* and *per-*; for example, *oxygen* and *chlorine* form four acids, as follows:—

Hypochlorous acid, $ClHO$.	*Chloric acid,*..........$ClHO_3$.
Chlorous acid,...... $ClHO_2$.	*Perchloric acid,*......$ClHO_4$.

The use of the prefix "*per-*" is generally limited to the compound containing the largest known proportion of *oxygen.*

Some acids do not contain *oxygen,* but have *sulphur* instead; these have a prefix *sulph-* or *sulpho-*. Thus, a union of *sulphur, arsenic,* and *hydrogen* produces *sulpharsenic acid;* while one of *sulphur, hydrogen,* and *tin* produces *sulphostannic acid.* In these acids, as in the case of the oxygen acids, the less or greater proportions of sulphur present are denoted respectively by the terminations "*ous*" and "*ic.*"

When an acid is formed by the binary compound of hydrogen and another element, it takes the prefix *hydr-* or *hydro-*; in this case the terminations "*ic*" and "*ous*" are not needed, since no element forms more than one acid with hydrogen.

* Some acids which do not contain oxygen, as *hydrochloric, hydrobroic,* &c., nevertheless terminate in "*ic.*"

When a proportion of water is abstracted from any oxygen acid the residue is called an anhydride; thus—

H_2SO_4 *Sulphuric Acid.*	—	H_2O *Water.*	=	SO_3. *Sulphuric Anhydride.*
$2\,NO_3H$ *Nitric Acid.*	—	H_2O *Water.*	=	N_2O_5. *Nitric Anhydride.*

33. Base.—Other binary and ternary compounds which never become acids, but which under all circumstances combine with the acids, and either neutralize them partially or entirely, are called *bases.*

34. Alkali.—Those bases which neutralize acids entirely are called *alkalies.*

The term *alkali* is of Arabic origin (*al*, the; *kali*, plant); and was given in the first instance to sodic carbonate, which was obtained from the ashes of plants. It is now extended to a large class of substances, which, like sodic carbonic, are soluble in water, possess an acrid, nauseous taste, restore the blue colour to vegetable infusions which have been turned red by an acid, and turn many vegetable blues green, as for instance, the solutions of blue cabbage or of litmus. Alkalies also turn vegetable yellows, as those of rhubarb or turmeric, brown; but this test is not so delicate as that of its power of neutralizing an acid by restoring the blue colour to litmus paper which has been feebly reddened by an acid.

The *bases* then may be divided into two classes—*First*, The oxides of metals; *Second*, Compounds of metals with a certain substance called *hydroxyl* (H_2O_2); these compounds are called *hydrates.* Some of these oxides and hydrates possess the power of entirely neutralizing or destroying the characters of the *acids;* these are called *alkalies.* To the latter class must be added *ammonia* (H_3N), which, though neither oxide nor hydrate, forms nevertheless a powerful base and alkali; it is a type of a numerous class of bodies met with in organic chemistry.

The first class of bases, the '*oxides,*' are named strictly in accordance with the law given for the naming of binary compounds, as *baric oxide, magnesic oxide,* and so on. When two oxides are formed, they are distinguished as before by the terminations "*ous*" and "*ic.*" When more than two oxides are formed, the number of atoms of *oxygen* to those of *metal* are signified by prefixes, as *di-*, *tri-*, *tetra-*; or in some cases the first oxide is called the *monoxide;* the second, the *binoxide.* In all cases the highest compound in a series often receives the prefix *per-*.

When the elements combine together in the proportion of two atoms of the one to three atoms of the other, the prefix "*sesqui-*" (one and a half) is employed, as *sesquioxide of iron,* (Fe_2O_3).

The oxides of the following metals, whose names end in "*um*" or "*ium,*" are also known commonly by the following names:—

METALS—	OXIDES—
Potassium,	*Potassa.*
Calcium,	*Lime.*
Strontium,	*Strontia.*
Barium,	*Baryta.*
Aluminium,	*Alumina.*
Magnesium,	*Magnesia.*
Glucinum,	*Glucina.*
Zirconium,	*Zirconia.*

The bases of the second class or hydrates have simply the name of the metal or positive element with the termination "*ous*" or "*ic*" before the name *hydrate* formed from *hydroxyl, e.g.*

Potassic hydrate,	KHO.
Sodic hydrate,	NaHO.
Zincic hydrate,	Zn2HO.

and so on.

35. Salts.—When an acid and a base unite together a *salt* is produced. If the acid contained *sulphur* instead of *oxygen,* it is called a *sulphosalt.* If the acid contained neither sulphur nor oxygen, it is called a *haloid salt* (ἅλς, *sea-salt*); but if the acid be one containing oxygen it is a salt proper.

The nomenclature of the salts is exceedingly simple: those which are formed from acids terminating in "*ous*" receive a termination "*ite*"; while those formed by an acid ending in "*ic*" are made to terminate in "*ate.*" Thus, a salt formed by *sulphurous* or *nitrous acids,* would be named a *sulphite* or *nitrite;* while those formed by *sulphuric* or *nitric acid,* would receive the name of a *sulphate* or *nitrate.* Any prefixes added to the acid would also be carried on to the salt: thus, a salt formed by *hyposulphurous acid* would be called a *hyposulphite;* while one formed by *perchloric acid,* would be called a *perchlorate,* and so on.

36. Symbolic Notation.—It has been already remarked that chemists, for the sake of convenience, have adopted a principle of notation by symbols. It is a kind of short-hand, which greatly abridges the labour of description, and enables the changes and reactions that take place to be briefly and clearly represented, even when they are of a most complicated character. For the elements, the initial letter of their Latin name is made use of; and, in the case of two or more elements commencing with the same letter, the single initial is reserved for the earliest discovered or most important element; the others being distinguished by the addition of a small letter to the initial one. Further, these symbols represent not only the element, but one *atom* or combining proportion of the element. When more than one atom or proportion of the element is intended to be represented, it is done either by writing the number before the symbol of the element; or, now almost universally, by writing a small figure to the right of the symbol, and below the line. Thus, H would stand for one atom of *hydrogen;* $2H$, or, more correctly, H_2, for two atoms; H_3, for three atoms, and so on.

37. Chemical Formulæ.—When a compound body is intended to be represented, the symbols of its constituent elements are simply placed in juxtaposition. Thus, HCl is the symbol for *hydrochloric acid;* H_2O is the symbol for water; Ag_2O for *argentic oxide;* H_2SO_4 for *sulphuric* *acid.* Such a group of two or more symbols is called a *chemical formula.* We may then define a *chemical formula* to be "the expression, by symbols and numbers, of the composition of a chemical compound."

In the *ordinary* symbolic language, the symbol of the most electro-positive element is placed first in the formula. Thus, *water* is written H_2O, not OH_2; *ammonia* is H_3N, not NH_3, and so on.

When a comma is used to separate the members of a formula, these members are represented as united chemically, and a more intimate union is supposed to exist than when the members are separated by a period.

A large figure placed before a symbol multiplies every symbol and figure up to the next comma or + sign; *e.g.* $3BaN_2O_6$ indicates $Ba_3N_6O_{18}$, or three parts of *nitrate of baryta.* But *nitrate of baryta* may also be written $Ba,2NO_3$, indicating that one atom or proportion of *barium* is united with two atoms or proportions of NO_3, in which case, to express three atoms of the salt, a bracket must be employed. Thus, $3(Ba,2NO_3)$, as otherwise, on account of the comma, the 3 would have multiplied only the *Ba*, and not the $2NO_3$. Brackets are not now much employed, but, when they are, the effect of a number before them is to multiply all the terms within them.

The + sign should never be used to connect together the constituents of the same compound, but only when two different bodies are added to or mixed with each other.

The − sign indicates abstraction, but it is seldom employed.

The formula for sulphate of zinc is sometimes written $ZnSO_4$, and sometimes ZnO,SO_3. In the first, it is merely intended to represent the actual composition of the salt, without in any way asserting how its component atoms are linked together. Such a formula is called an *empirical formula,* and may be regarded as a simple statement of fact involving no theory whatever. In the second case, the formula asserts that one atom of oxygen is united with one atom of zinc to form oxide of zinc, which again unites itself to one part of sulphuric anhydride (SO_3) to form *sulphate of oxide of zinc*, or more briefly, *sulphate of zinc*, or *zincic sulphate*. Such a formula is called a *rational formula*, because it attempts to account for the way in which the constituent atoms are grouped together. Such a formula may be, and probably is, true in the majority of cases; but it must ever be borne in mind that it is only theory.

38. Chemical Equations.—The changes or reactions which take place under the influence of chemical action may be represented in two ways—either by means of a diagram, or by an equation.

If we mix together solutions of *common salt (sodic chloride)* and *argentic nitrate* in proper proportions *(sodic chloride, $NaCl = 58{\cdot}5$; argentic nitrate, $AgNO_3 = 170$)*, the *sodium* and the *silver* change places, and *sodium nitrate* and *argentic chloride* are formed. This double decomposition, as it is called, may be represented by a diagram, as follows—

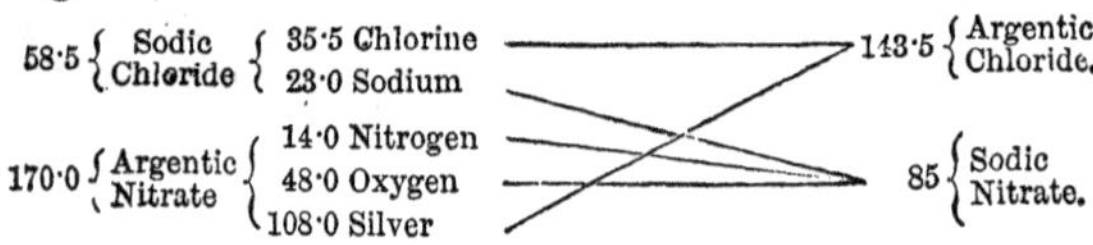

This change may be, however, better and more simply represented by means of an equation, in which the substances as they exist before the reaction are placed on the left hand side, while those which are formed by the reaction are on the right. Thus—

$NaCl$	+	$AgNO_3$	=	$AgCl$	+	$NaNO_3$.
Sodic chloride.		Argentic nitrate.		Argentic chloride.		Sodic nitrate.

Or

Zn.	+	Cl_2	=	$ZnCl_2$.
Zinc.		Chlorine.		Zincic chloride.

$2HCl$	+	Na_2	=	$2NaCl$	+	H_2.
Hydrochloric acid.		Sodium.		Sodic chloride.		Hydrogen.

The sign = in a chemical equation does not mean "equal to," but is to be regarded rather in the sense of "*yields*" or "*produces,*" or "*is converted into.*" Thus—

Zn	+	H_2SO_4	=	$ZnSO_4$	+	H_2.
Zinc.		Sulphuric acid	produces	Zincic sulphate.		Hydrogen.

CHAPTER VI.*

Atomicity of Elements—Classification according to Atomicity—Graphic Notation—Simple and Compound Radicals—Definition of a Compound Radical.

39. THE combining or atomic weights of the elements were formerly called equivalent numbers, as these proportions were considered equal to each other in chemical combination; but more accurate knowledge shows us to have been in error. We now find that the atoms of the different elements are by no means equal in chemical combination. For instance, sulphuric acid has the following formula, H_2SO_4; if we now put to it sodium or potassium, we find that for every atom or equivalent of metal taken up, one atom or equivalent of hydrogen is set free, as shown by the following equations:—

(1.) $H_2SO_4 + K = KHSO_4 + H$.
(2.) $H_2SO_4 + Na = NaHSO_4 + H$.

While if zinc or iron be added to the sulphuric acid, for every atom or equivalent of metal taken up two atoms of hydrogen are set free, as follows:—

(3.) $H_2SO_4 + Zn = ZnSO_4 + H_2$.
(4.) $H_2SO_4 + Fe = FeSO_4 + H_2$.

* The whole of this chapter is, by Professor Frankland's kind permission, taken from his *Lecture Notes for Chemical Students,* in many parts the passages being copied *verbatim.*

So that, while the atoms of K and Na are capable of replacing only one atom of H in a chemical combination, those of Zn and Fe can replace two atoms of H; clearly, then, the atoms of K or Na are not equivalent to either those of Zn or Fe.

If we regard the following series of compounds as typical, viz.:—

Hydrochloric Acid,	HCl.
Water,	H_2O.
Ammonia,	H_3N.
Marsh Gas,	H_4C.

we shall see that atoms of Cl, O, N, and C require, or are capable of combining with, 1, 2, 3, and 4 atoms of H respectively, or, in other words, that oxygen has twice, nitrogen three times, and carbon four times, the atom fixing power, as far as hydrogen is concerned, that chlorine has.

Again, while hydrogen and chlorine combine together atom to atom, as in hydrochloric acid (HCl), other elements are capable of taking up different numbers of atoms of chlorine, as in the following series:—

Hydrochloric Acid,	HCl.
Baric Chloride,	$BaCl_2$.
Auric, ,,	$AuCl_3$.
Platinic ,,	$PtCl_4$.
Antimonic ,,	$SbCl_5$.

Thus, one atom of *barium* combines with two atoms of *chlorine*, and is therefore equal to (in combining power) two atoms of *hydrogen;* in like manner one atom of gold may be regarded as equal to three atoms of *hydrogen*, one of *platinum* to four atoms, and one of *antimony* to five atoms of *hydrogen*.

From this we see clearly that the equivalents of elements are not necessarily their atomic weights.

40. Definition of Atomicity.—By *equivalence* or *quantivalence* or *atomicity* of an element, we mean the number of atoms of hydrogen (and therefore also of chlorine) which it is capable of replacing in a compound.

41. Monads.—Those elements which, in combination, only replace one atom of hydrogen or chlorine are called *monads, monequivalent,* or *univalent elements.* The chief are, Na, K, Ag, Br, I, F, &c. The compounds they form are, *NaCl* (*sodic chloride*), *KCl* (*potassic chloride*), *AgCl* (*argentic chloride*), *HBr* (*hydrobromic acid*), *HI* (Hydriodic acid.

42. Dyads.—Elements which, like *oxygen, barium, calcium,* &c., are capable of replacing two atoms of *hydrogen* in combination, are called *dyads, diequivalent* or *divalent elements.*

43. Triads.—Those which, like *gold* and *boron,* can replace three atoms of *hydrogen,* are called *triads, triequivalent* or *trivalent elements.*

44. Tetrads.—*Tetrads, tetriquivalent* or *tetravalent elements,* are those which, like *carbon, tin, silicon, platinum, lead,* &c., can replace *four* atoms of *hydrogen.*

45. Pentads.—*Pentads, pentequivalent* or *pentavalent elements,* are those which, like *nitrogen, phosphorus,* &c., can replace *five* atoms of *hydrogen.*

46. Hexads.—*Hexads, hexequivalent* or *hexavalent elements* are those which replace *six* atoms of *hydrogen,* e.g. *sulphur, iron, cobalt, nickel,* &c.

The following classification of the atomicity of elements is taken from Dr. Frankland's *Lecture Notes for Chemical Students,* page 32. It is not quite in accordance with that given by the late Dr. Miller or by other authors; but it is, we think, the one most easily understood and proved.

MONADS.	DYADS.	TRIADS.	TETRADS.	PENTADS.	HEXADS.
1st Section. **HYDROGEN.**	1st Section. **Oxygen.**	1st Section. **BORON.**	1st Section. **CARBON.** **SILICON.** TIN. TITANIUM.	1st Section. **NITROGEN.** **PHOSPHORUS.** ARSENIC. ANTIMONY. BISMUTH.	1st Section. **Sulphur.** **Selenium.** Tellurium.
2nd Section. **Fluorine.** **Chlorine.** **Bromine.** **Iodine.**	2nd Section. BARIUM. STRONTIUM. CALCIUM. MAGNESIUM. ZINC.	2nd Section. GOLD.	2nd Section. THORINUM. NIOBIUM. TANTALUM. ZIRCONIUM. ALUMINIUM.		2nd Section. TUNGSTEN. VANADIUM. MOLYBDENUM.
3rd Section. CÆSIUM. RUBIDIUM. POTASSIUM. SODIUM. LITHIUM.	3rd Section. DIDYMIUM. LANTHANUM. YTTRIUM. GLUCINUM.		3rd Section. PLATINUM. PALLADIUM.		3rd Section. OSMIUM. IRIDIUM. RUTHENIUM. RHODIUM.
4th Section. THALLIUM. SILVER.	4th Section. CADMIUM. MERCURY. COPPER.		4th Section. LEAD.		4th Section. CHROMIUM. MANGANESE. IRON. COBALT. NICKEL. URANIUM. CERIUM.

In the above table the non-metals are printed in thick type, and the metals are put in ordinary type, whilst the positive or basylous elements are printed in capitals, and the negative or chlorous elements in small letters.

In the symbolic notation the atomicity of the elements

is marked by accents or equivalence marks, and Roman numerals placed above and to the right of the symbol; thus, H^{i}, O^{ii}, B^{iii}, C^{iv}, N^{v}, S^{vi}.

47. Graphic Notation.—In the *graphic notation* the elements are represented by a ball having the symbol of the element inside of it, and having as many rods or pegs proceeding from it as mark its atomicity; thus hydrogen, (H); oxygen, -(O)-; boron, (Bo); carbon, -(C)-; nitrogen, (N); sulphur, -(S)-

These rods or pegs are regarded as forming points of attachment for other bonds and pegs of other atoms either of the same or different elements. Thus, an atom of hydrogen cannot exist independently, as (H), there being a bond unsatisfied; but two atoms of hydrogen, forming a molecule,* can exist, the bonds of each mutually satisfying each other, thus, (H)—(H). It is yet a disputed point whether the bonds of the same atom can satisfy each other; thus, whether the atom of oxygen can exist alone in this manner, (O), or whether we must regard it as a molecule, thus, (O)═(O). Dr. Frankland seems to regard it as possible that the bonds can satisfy each other; and in a large number of compounds we can explain the facts in no other way than that they do satisfy each other.

Hydrochloric acid may be thus graphically represented, (H)—(Cl).

Water, thus (H)—(O)—(H).

* A molecule is regarded as the smallest portion of an element or compound which is capable of a separate chemical existence. It may consist of two or more atoms.

Magnesic oxide, thus, (O)=(Mg), (both elements being *dyads*).

Ammonia, thus, (H)—(N)—(H), with a third (H) bonded to (N); and so on.

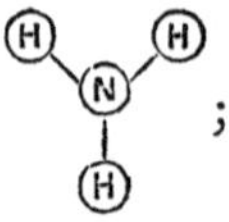

The element *nitrogen* appears as a *triad* in ammonia,

$\mathbf{N}H_3$ (three (H) bonded to (N)) and as a *pentad* in *ammonic chloride*,

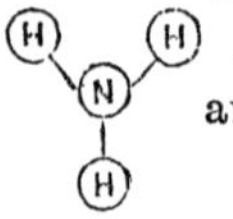

$\mathbf{N}^{v}H_4Cl$. (four (H) and one (Cl) bonded to (N))

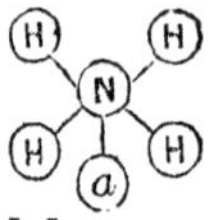

Again, *sulphur* appears as—

Dyad in *Hydric sulphide*, $\mathbf{S}^{ii}H_2$ (H)—(S)—(H)

Tetrad in *Sulphurous anhydride*, $\mathbf{S}^{iv}O_2$ (O)=(S)=(O)

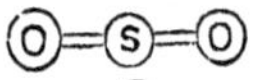

Hexad in *Sulphuric anhydride*, $\mathbf{S}^{vi}O_3$ (O)=(S)=(O), with a third (O) double-bonded to (S)

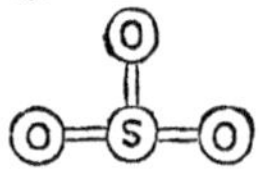

It is then a law, to which there are no real exceptions, (the apparent ones admit of a simple explanation), that though the equivalence of an element may vary, it does so always by the addition or subtraction of an even number—*i.e.*, a pentad element may become a triad, or even a monad, but can never, under any circumstances, become a dyad, tetrad, or hexad; so again, a hexad may in some combinations appear as a tetrad or a dyad, as in the cases of sulphur quoted above, but can never be a monad, triad, or pentad. In other words, an element having an even equivalence cannot ever become odd-equivalent; nor can an element which has odd equivalence ever possess even equivalence.

48. Perissads and Artiads.—On this account Dr. Odling

has proposed to divide the elements into two classes, "*perissads*" and "*artiads*," (from περισσός, uneven, odd; and ἄρτιος, even), or *perissequivalent*, and *artiequivalent* elements.

Dr. Frankland's explanation of this diversity of equivalence of the same element is this—"that one or more pairs of bonds belonging to one atom of the same element can unite, and having saturated each other, become, as it were, latent." Thus the pentad nitrogen becomes a triad when one pair of its bonds becomes latent; and a monad when two pairs, by combination with each other, are rendered latent. These conditions are represented graphically, thus—

Pentad. Triad. Monad.

(N) (N) (N)

And so in the case of sulphur—

Hexad. Tetrad. Dyad.

(S) (S) (S)

49. Absolute, Active, and Latent Atomicity.—Dr. Frankland, on this hypothesis, regards all the bonds which an element possesses as its *absolute atomicity;* those which are concerned in linking it with the other elements of a compound as its *active atomicity;* and the number of bonds which are united together as its *latent atomicity.* So that the sum of its latent and active atomicities must always be equal to its absolute atomicity.

From these facts on equivalence we deduce two important conclusions, viz.:—

1. That the sum of the bonds of a molecule must always be an even number, since the atoms are bound together by the bonds which constitute the equivalence of their component atoms, and that no bond can be left disengaged or unsatisfied. So that a formula which possesses an uneven number of bonds or units of chemical affinity, cannot possibly represent a molecule.

2. No portion of a molecule can be removed without leaving one or more of the bonds of the molecule unsatisfied, the substances left are therefore incapable of a separate existence, *e. g.*, a molecule of marsh gas is—

Symbolic. Graphic.

$C^{iv}H_4$, 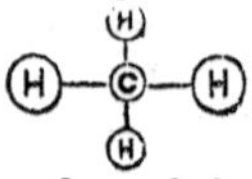 ; if we take away one atom of hydrogen, we have left a molecule of

Methyl, $C^{iv}H_3$ 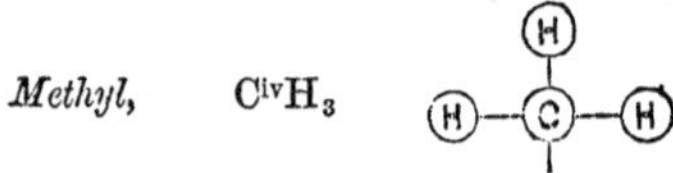

which has one bond unsatisfied; and is, therefore, ready to unite with any monad atom, or with any molecule simple or compound which has one unsatisfied bond. Now, if we take from this another atom of *hydrogen* we leave a molecule of

Methylene, $C^{iv}H_2$ 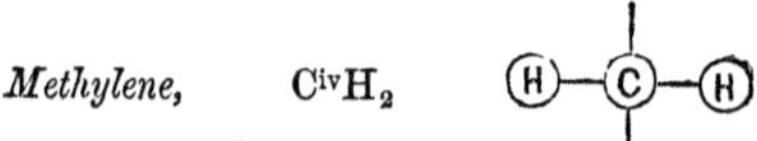

which, from its having two bonds unsatisfied, may be regarded as a *dyad compound molecule*; if, further, from this molecule of *methylene* we abstract another atom of *hydrogen* we shall have a molecule of

Formyl, $C^{iv}H$

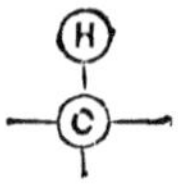

which may be viewed as a *triad compound molecule;* and, if we still further abstract this remaining atom of *hydrogen*, we shall leave the *tetrad elementary atom* of carbon C^{iv}  , having four bonds unsatisfied.

Every atom or molecule which, from its having bonds unsatisfied, is ready to enter into combination, is called a *radical;* if it is elementary or simple in its nature, it is called a *simple radical;* if it is compound, it is called a *compound radical.*

50. Definition of Compound Radical.—When two or more *simple radicals* (*i.e.*, elements) combine, they either form a *compound radical* or *salt.*

They form a compound radical if the substance produced is capable of entering further into chemical composition or combinations, without itself suffering decomposition. Thus—

$$\underset{\text{Potassium.}}{K_2} + \underset{\text{Hydroxyl.}}{2O''H} = O''KH + O''KH.$$

In the above series of combinations of C and H, *methyl, methylene,* and *formyl,* are all of them *compound radicals,* and are respectively regarded as *monad, dyad,* and *triad,* according to the number of bonds of attachment which remain unsatisfied. All compound radicals have their equivalence marked on the same principle The three in question may be written thus—*methyl* (CH_3);* *methylene* $(CH_2)''$; *formyl* $(CH)'''$.

Sometimes it is necessary to give names to compound radicals, and to substitute for their formulæ a single symbol, *e.g.*, the radical *methyl* $C^{iv}H_3$ is represented by symbol Me; so the radical $O''H$ is called hydroxyl, and is represented by symbol Ho.

The following are the names and formulæ of the chief inorganic compound radicals recognized by chemists :—

	Molecular formulæ.	Atomic formulæ.	Abbreviated atomic formulæ.
Hydroxyl, ...	$(HO)_2$	HO	Ho.
Hydrosulphyl,............ ...	$(HS)_2$	HS	Hs.
Ammonium, ...	$(NH_4)_2$	NH_4	Am.
Ammonoxyl,.................	$(NH_4O)_2$	NH_4O	Amo.
Amidogen, ...	$(NH_2)_2$	NH_2	Ad.

* It is customary to omit the equivalence marks to monad radicals, both simple and compound.

Besides these, certain compounds which metals form with oxygen are regarded as compound radicals, as—

	Molecular formulæ.	Atomic formulæ.	Abbreviated atomic formulæ.	
Potassoxyl, ...	$(KO)_2$	KO	Ko	monad.
Sodoxyl, ...	$(NaO)_2$	NaO	Nao	
Zincoxyl, ...	(ZnO_2)	$\begin{cases} O \\ Zn'' \\ O \end{cases}$	Zno″.	
Cuproxyl,.........	(CuO_2)	$\begin{cases} O \\ Cu \\ O \end{cases}$	Cuo″.	

And so on.

The essential character of these last compound radicals is that the whole of the oxygen they contain is united with the metal by one bond only of each oxygen atom, as seen in the following graphic formulæ :—

Hydroxyl,........................... –(O)—(H)

Potassoxyl, ... –(O)—(K)

Zincoxyl, ... –(O)—(Zn)—(O)–

The metal thus becomes linked to other elements by these dyad atoms of oxygen. The functions of such compound radicals will be sufficiently evident from the following examples of compounds into which they enter, and in which their position is marked by dotted lines :—

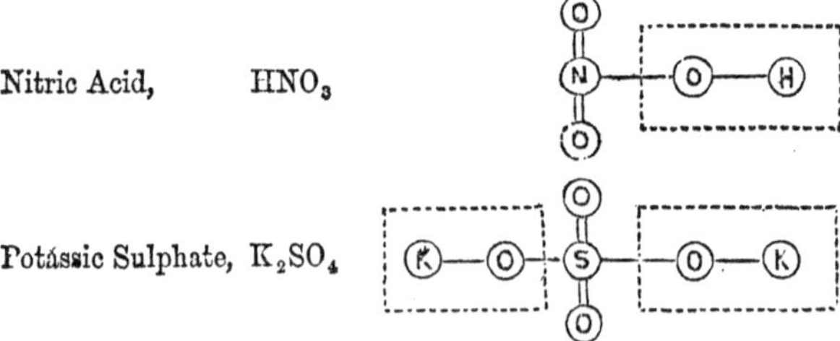

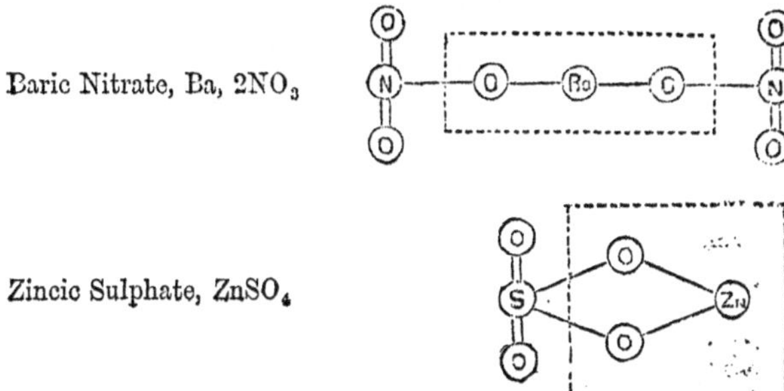

It is not necessary to name all these metallic compound radicals. The only point of importance is their abbreviated notation in which the small letter *o* is attached to symbol of the metal—the atomicity of the radical being marked in the usual manner.

51. Use of Thick Type.—The system of graphic formulæ attempts to show how each unit of force of each of the atoms is expended; but the same thing can be shown by means of symbols. For this purpose the first symbol of a formula should be that of the element which is directly united with all the active bonds of the other elements or compound radicals which follow it upon the same line: thus, the formula $\mathbf{S}O_2Ho_2$ (sulphuric acid) signifies that the hexad atom of sulphur is combined with the four bonds of the two atoms of *oxygen*, and also with the two bonds of the two atoms of *hydroxyl.**

When the element which is thus placed first has more than one bond, it will be printed in thick type, and, as a rule, the element having the greatest number of bonds will occupy this position. Thus—

* Such a formula is called a *rational formula*, as it attempts to account for the way in which the elements are united together. When, however, a formula simply gives the composition of a substance without regard to the arrangement of its components, it is called an *empirical formula.*

	Empirical Formulæ.	Rational Formulæ.
Water, ...	H_2O	**O**H_2.
Sulphuric acid, ...	H_2SO_4	**S**O_2Ho_2.
Nitric acid, ...	HNO_3	**N**O_2Ho.
Microcosmic salt, ...	NaH_4NHPO_4	**P**OHoAmoNao.

52. Use of Bracket.—These atoms in thick type form the central or governing atoms around which the others are grouped, to indicate which a bracket is used, as in the following three examples :—

$$1.\ \left\{\begin{array}{l} CH_3. \\ CH_3. \end{array}\right. \qquad 2.\ \left\{\begin{array}{l} CH_3. \\ O. \\ CH_3. \end{array}\right. \qquad 3.\ \left.\begin{array}{r} COO. \\ Ba. \\ COO. \end{array}\right\}$$

"The formula No. 1 signifies that two atoms of carbon "are directly united with each other; No. 2, that two "atoms of *carbon* are linked, as it were, together by an "atom of *oxygen*, the latter being united to both carbon "atoms; whilst in like manner No. 3 expresses the fact "that one atom of *oxygen* in the formula of the upper line "is linked to another atom of *oxygen* in the formula of the "lower line by an atom of *barium*." So, again, ferric chloride, the empirical formula for which is Fe_2Cl_6, can be represented by the thick letter symbols in two ways, thus—

Graphic

$$\text{Ferric chloride, } \left\{\begin{array}{l} 'Fe'''Cl_3 \\ 'Fe'''Cl_3 \end{array}\right.$$
or
$$'''Fe_2'''Cl_3$$

indicating that the two atoms of iron (each being hexad) are linked to each other by three of their bonds, leaving their other three bonds available for their union, each with three atoms of chlorine.

The equivalence marks are not inserted with the monad elements; nor in the case of oxygen, which is invariably dyad; nor with *carbon*, unless its equivalence be reduced to ″. The marks are also omitted in cases where there can be no difficulty in assigning the equivalence of the elements.

It will be thus seen that this method of symbolic notation, by means of equivalence marks and thick letters, expresses exactly the same thing as the graphic notation, which is somewhat cumbrous for general use; but still the graphic system, especially if supplemented by Hofmann's glyptic method, is of incalculable use in the lecture theatre, in bringing vividly before a class, and making easy of comprehension, the changes that take place in even the most simple reactions.

The objection made to the graphic and glyptic, and also to the symbolic formulæ of Dr. Frankland, viz., that it is a purely theoretical attempt to give a picture of the physical arrangement of atoms, is, I think, repudiated by every teacher, who has made a fair and impartial trial of the methods.

In the remainder of this little work, when treating of the elements, their preparation, properties, and combinations, I shall express all reactions in equations, using Dr. Frankland's method of symbolic notation; but, for the convenience of those who do not understand or approve of that method, and for the sake of comparison with the many standard works which do not use it, such as those of Miller, Roscoe, Fownes, Galloway, &c., I shall add, in brackets thus [ ], the equation in the old method of notation. I shall also, wherever I think it useful, give the graphic formulæ.

CHAPTER VII.

Hydrogen—its History—Distribution and Natural History—Preparation—Properties—Combinations.

53. Hydrogen.—Symbol H.—Atomic weight = 1. Atomic volume, □. Density = 1. Specific gravity = 0·0692. 1 litre weighs 1 crith. Molecular weight, 2. Molecular volume, □□. Atomicity, ′, being the standard of comparison.

54. History.—The element *hydrogen* was first mentioned in the sixteenth century by Paracelsus, but very little was known of it in any way. According to Faraday, it seems to have been regarded in the 16th century as synonymous with the "*phlogiston*" of Beecher and Stabil, from its burning so eagerly in the air. It was afterwards called inflammable air, for the same reason, and was confounded with several of its compounds. In 1766, it was first obtained in a pure state, and its properties thoroughly examined; and soon afterwards it was named *hydrogen* (ὕδωρ, water; γεννειν, to generate) by Lavoisier, from its being found that its combination with *oxygen* produced *water.*

55. Distribution and Natural History.—*Hydrogen* is very rarely found free in nature, but exists very largely in combination, both in the inorganic and organic kingdoms. In the former it is found as a constituent of various acids in combination, as the hydrochloric, hydrobromic, hydriodic, and hydrofluoric acids. It is a constituent of certain liquids, as water, naphtha, petroleum; and also of certain solids, as in the salts of ammonia.

In the organic kingdom, it is found largely entering into the composition of both animals and vegetables, in the forms of *water* (H_2O) and *ammonia* (H_3N).

56. Preparation.—It is most generally obtained from water, by the action of substances on the water, which are capable of removing *oxygen* from it, and setting free the *hydrogen.*

(1.) Thus, if pieces of the following metals—potassium, sodium, lithium, barium, strontium, calcium, magnesium,—are added to water, they will decompose it, and set free the whole, or a portion of the *hydrogen.*

The first three—K, Na, and Li—being *monads,* displace only one part of the hydrogen contained in water, —the reaction which takes place being represented by the following equation :—

$$\underset{\text{Water.}}{2\mathbf{O}H_2} + \underset{\text{Sodium.}}{Na_2} = \underset{\text{Sodic hydrate.}}{2\mathbf{O}NaH} + \underset{\text{Hydrogen.}}{H_2}$$

The last four—Ba, Sr, Ca, and Mg—are *dyads,* and therefore displace two proportions of hydrogen, as shown in the following equation :—

$$\underset{\text{Water.}}{2\mathbf{O}H_2} + \underset{\text{Calcium.}}{Ca} = \underset{\text{Calcic hydrate.}}{2\mathbf{Ca}''Ho_2} + \underset{\text{Hydrogen.}}{H_2}$$

These last four decompose water with very much less energy than the three first named. The metal Mg requires the water to be heated to a temperature of 55° C. (122° F.) before it is able to effect its decomposition.

This first method of obtaining *hydrogen* is not the one practically followed, the expense and difficulty being too great. Potassium acts on the water with so great violence that it is impossible to collect the hydrogen.

EXP. 23.—Throw a piece of potassium about the size of a small nut upon the surface of a plate or flat glass dish, filled with water, part of the water is immediately decomposed; each atom of potassium displaces half the hydrogen of an atom of water, and potassic hydrate is formed.

$$\underset{\text{Water.}}{2\mathbf{O}H_2} + \underset{\text{Potassium.}}{K_2} = \underset{\text{Potassic hydrate.}}{2\mathbf{O}KH} + \underset{\text{Hydrogen.}}{H_2}$$

$$[2H_2O + K_2 = 2KHO + H_2]$$

The escaped hydrogen takes fire from the heat resulting from the intensity of the chemical combination, and carrying with it a small portion of the volatilized metal, burns with a reddish violet flame; the metal melts and swims about rapidly upon the water, and finally disappears with an explosive burst of violence, as the globule of melted potassic hydrate which is formed during its oxidation becomes sufficiently cool to come into contact with the water.

The action of *sodium* on the water is the same as that of *potassium*, *sodic hydrate* is formed, which being soluble, is immediately dissolved, and *hydrogen* is set free, as in the following equation:—

$$2\mathbf{O}H_2 + Na_2 = 2\mathbf{O}NaH + H_2$$

Water. Sodium. Sodic Hydrate. Hydrogen.

$$[2H_2O + Na_2 = 2NaHO + H_2]$$

Sodium does not, however, act with the same energy as potassium, and consequently the evolved gas does not take fire, unless the water be warm, or the sodium be confined to one place, by wrapping it in a piece of filtering paper, previous to throwing it on the water. Advantage is taken of this to collect the hydrogen evolved, as in the following experiment—

Exp. 24.—Take a piece of sodium as large as a pea, place it securely in a small wire-gauze box, or wrap it tightly in wire gauze firmly fastened to the end of a wire (A, fig. 8), and hold it under an inverted cylinder or gas jar (B) filled with water, and standing on a bee-hive shelf. The bubbles of hydrogen gas ascend in the cylinder and displace the water. When the cylinder is full, it may be put on one side for future examination of the properties of the gas. This mode of obtaining hydrogen is too expensive, too troublesome, and too dangerous, to be used otherwise than as an experiment of demonstration.

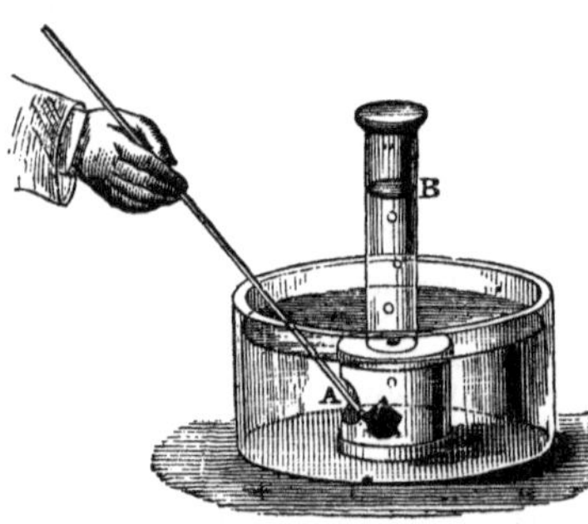

Fig. 8.

(2.) A second method of obtaining *hydrogen*, is by the action of *sodium* on dry hydrochloric acid. The reaction which takes place is represented by the following equation–

$2HCl$	$+$	Na_2	$=$	$2NaCl$	$+$	H_2
Hydrochloric acid.		Sodium.		Sodic chloride.		Hydrogen.

(3.) Hydrogen may be prepared very simply and cheaply in large quantities by passing steam over iron filings, or turnings, or even iron nails, maintained at a red heat. The hydrogen so prepared contains a deal of moisture, and therefore if it is required to be dry it must be passed through tubes containing chloride of calcium. The reaction is as follows—

Fe_3	$+$	$4OH_2$	$=$	$^{iv}(Fe_3)^{viii}O_4$	$+$	$4H_2$
Iron.		Water.		Triferric tetroxide or magnetic oxide of iron.		
[$3Fe$	$+$	$4H_2O$	$=$	Fe_3O_4	$+$	H_8

The graphic formula for $^{iv}(Fe)^{viii}O_4$ is

O=Fe—Fe—Fe=O (with O bonded to each pair of adjacent Fe above)

If a small quantity of hydrogen only be required, or it be merely wanted to demonstrate this method of obtaining it, the arrangement shown in fig. 9 may be adopted.

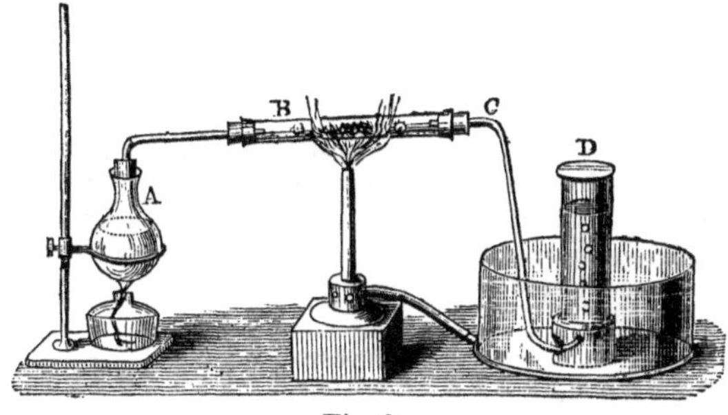

Fig. 9.

Exp. 25.—A is a flask containing water boiling briskly by the aid of a Bunsen burner or spirit lamp, and connected by a bent tube with a piece of combustion tube B, from 9 to 15 inches long, and about ⅝ of an inch in diameter, in which is placed about an ounce of iron filings, which are heated to redness by means of a Bunsen burner; C is the delivery tube, and D the receiver, standing inverted, over a pneumatic trough.

When a large quantity of hydrogen is required, an arrangement similar to that shown in fig. 10 must be made, in which the

Fig. 10.

glass combustion tube is replaced by an iron gun barrel, or, better still, by a Berlin porcelain combustion tube; and the source of heat, either a fire-clay furnace, heated with charcoal, as shown in the figure, or a gas combustion furnace especially formed for tube operations. The U-tube filled with calcic chloride is simply for drying the gas.

One litre of water converted into steam will yield about 1,240 litres, or 1·24 cubic metre of hydrogen.

(4.) The most common method of preparing *hydrogen* is by the action of metals on water in presence of an acid; the metal most commonly employed is zinc, and the acid, sulphuric; but iron may and sometimes does take the place of zinc, and hydrochloric acid may replace in like manner the sulphuric.

Zinc and sulphuric acid is, however, almost universally employed on account of its greater cheapness and also that it yields the gas purer than either of the other combinations. The reactions that take place will be respectively represented in the following equations :—

(1.)	$\mathbf{S}O_2Ho_2$	+	Zn	=	$\mathbf{S}O_2Zno''$	+	H_2
	Sulphuric acid.		Zinc.		Zincic sulphate.		Hydrogen.
	[H_2SO_4	+	Zn	=	$ZnSO_4$	+	H_2]
(2.)	2HCl	+	Zn	=	$\mathbf{Zn}Cl_2$	+	H_2
	Hydrochloric acid.		Zinc.		Zincic chloride.		
	[2HCl	+	Zn	=	$ZnCl_2$	+	H_2]
(3.)	$\mathbf{S}O_2Ho_2$	+	Fe	=	$\mathbf{S}O_2Feo''$	+	H_2
					Ferrous sulphate.		
	[H_2SO_4	+	Fe	=	$FeSO_4$	+	H_2]
(4.)	2HCl	+	Fe	=	$\mathbf{Fe}''Cl_2$	+	H_2
					Ferrous chloride.		
	[2HCl	+	Fe	=	$FeCl_2$	+	H_2

The preparation of hydrogen from zinc and sulphuric acid is conducted as follows:—

Exp. 26.—A generating flask, A, fig. 11, having a capacity of about a pint, and also a somewhat wide mouth, is chosen; to this a good sound soft cork is fitted, through which pass two tubes—one, a short right-angled one, open at both ends; the other, a straight one, terminated by a funnel, called a thistle-headed funnel, which should pass down nearly to the bottom of the flask. By means of a piece of caoutchouc tubing, the right-angled tube may be attached to one, and therefore to any number, of washing bottles,* which may contain water or any reagent, through which it is advisable for the gas to pass, to separate any impurities which may be associated with it.

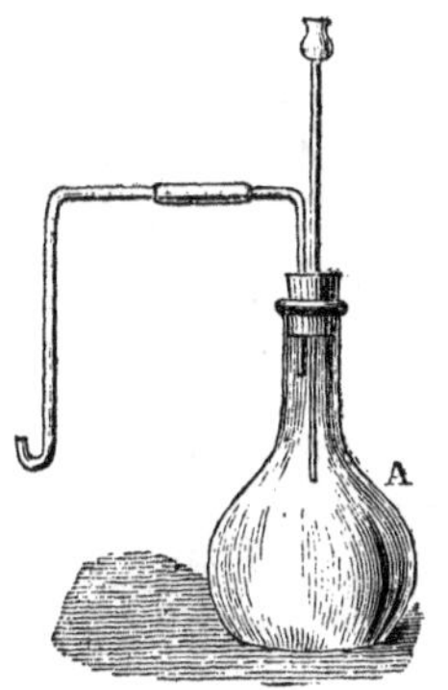

Fig. 11.

In the generating flask, A, should be placed some fragments

* Washing bottles are simply ordinary wide-mouthed bottles fitted with corks, through which pass two tubes, each bent at a

or cuttings of zinc, or better still, some granulated zinc, (which is made by melting the zinc in an iron ladle, and pouring it from a height of three or four feet into a pail of cold water), then pour into the gas bottle some dilute sulphuric acid, 1 of acid to 7 or 8 of water. (Great care is necessary in mixing the acid and water; the water should be placed in a thin glass beaker and the acid added gradually, great heat being evolved during the mixing. The dilute acid should not be used until it has cooled). The zinc should be covered to the depth of three or four inches. In a short time a brisk effervescence will be observed due to the escape of bubbles of hydrogen. The gas which at first comes off is, of course, contaminated with the atmospheric air contained in the generating flask, and is then dangerously explosive; after a short time, however, this will be all driven away, and the gas will be given off pure. To ascertain when this is the case, collect some of the gas from time to time in a small eprouvette, and apply a light to it; if it explodes or lights with a kind of whistling pop there is air mixed with it, but if it lights with a plain pop and keeps on burning at the mouth of the tube with a pale blue flame, it is pure.

Collect at the pneumatic trough as many jars or bottles of gas as may be required to illustrate its properties hereafter described. Should (as frequently happens) the evolution of the gas slacken before all the metal is dissolved, it can be quickened by the addition of a little more acid through the safety funnel.

The *rationale* of what takes place is this: The zinc drives out the hydrogen of the hydric sulphate or sulphuric acid and takes its place, forming zincic sulphate, and setting the hydrogen free; the zincic sulphate, being soluble, is dissolved by the water present, and can be obtained in white crystals from the residue left in the bottle, by simply evaporating the liquid.

(5.) *Zinc* or *iron*, when boiled with a strong solution of *potassic hydrate*, displaces the *hydrogen*. Thus—

$$2\mathbf{O}KH + Zn = \mathbf{Zn}Ko_2 + H_2$$

Potassic hydrate. Potassic zinc oxide.

$$[2KHO + Zn = ZnO + K_2O + H_2]$$

right angle, one consisting of a long and a short limb, the other of two short limbs; in the bottle is placed the liquid or other sub stance through which it is intended the gas should pass, and into which the long extremity of the tube C dips, the short tube D conveying the gas either to another washing bottle or to the delivery tube.

(6.) Hydrogen has been prepared on a large scale by the French chemists Deville and Debray, by passing steam over charcoal or coke, heated to a dull red heat. Thus—

$$\underset{\text{Water.}}{2\mathbf{O}H_2} + C^{iv} = \underset{\text{Carbonic anhydride.}}{\mathbf{C}^{iv}O_2} + H_2$$

$$[2H_2O + C = CO_2 + H_2]$$

The *carbonic anhydride* can be easily removed by passing the gas through a strong solution of *potassic hydrate.*

If, however, the heat be not strictly regulated—if it at all exceed the right degree—another gas, carbonic oxide (Co), which is highly poisonous, is also formed, and this gas cannot be removed.

(7.) By the electrolysis of water.

On plunging the poles or electrodes of a voltaic battery, consisting of two to four cells of Grove's, or not less than six large cells of Smee's arrangement,* into water slightly acidulated with sulphuric acid, *hydrogen* is given off at the negative electrode, or that connected with the zinc end of the battery; while *oxygen* is given off at the positive electrode, or that connected with the carbon or platinum end of the battery.

Fig. 12 gives a sketch of the arrangement employed.

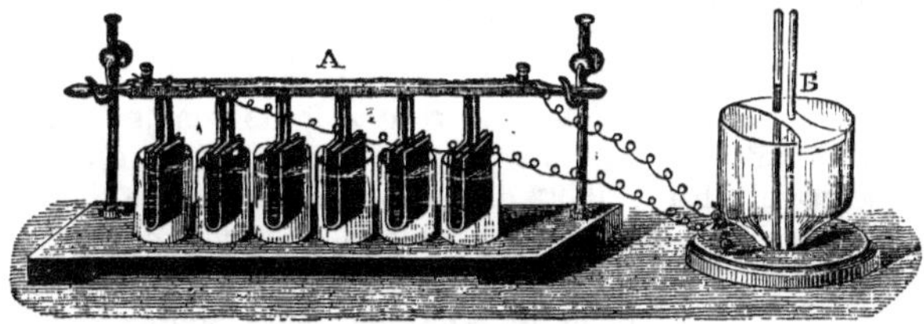

Fig. 12.

A is a battery consisting of six cells of Smee's, the poles

* The student should consult the volume on *Electricity* in this series, or some good manual of Natural Philosophy, for the difference between Grove's and Smee's batteries, and for the chemical action of the electric current.

or electrodes of which can be connected with the binding screws of a voltameter, B. These binding screws are connected by wires beneath with electrodes of platinum foil, over which are inverted glass tubes for the collection of the gases separately. It will be observed that the gas in one tube occupies twice the space which that in the other tube does. On testing the gases, that in the first tube will be found to possess all the characteristics of *hydrogen*, (see page), and that in the second tube those of *oxygen* (see page).

This method of procuring hydrogen is also valuable as an analytical experiment, proving that water is compounded of the two gases *hydrogen* and *oxygen* in the proportion by volume of 2 to 1; and as the weights of equal volume of H and O are as 1 to 16, it follows that in water they are related to each other by weight in the proportion of 1 to 8.

57. Properties.—Hydrogen is a gas, and as no amount of cold or pressure yet obtained has succeeded in condensing it, it is regarded as one of the *permanently elastic gases;* it is colourless, tasteless, and inodorous; it is very slightly soluble in water—fifty volumes of water dissolving a little more than one part of the gas; it will not support combustion, in the ordinary sense of the word;* but, in the presence of oxygen or of ordinary atmospheric air, is the most combustible body known. This is, in fact, its chief chemical peculiarity. The sole result of its combustion with O is water. It is the lightest body in nature, its specific gravity being ·0692 as compared with air, so that air is 14·4, or about 14½ times as heavy as hydrogen. Its great lightness has caused it to be taken as the standard to which the weight of all other gases, simple and compound, is referred.† It also possesses the greatest power of diffusion of all gases.

* For a full account of the true theory of combustion, see chapter , page .

† See paragraph 29, on the "Crith" and its uses.

(By *diffusion*, we are to understand "*the tendency of the particles of a gas to separate as far as they can from each other.*")*

Hydrogen is not poisonous, and may be breathed once or twice with impunity; but, if continued, it is fatal. This proves, therefore, that it cannot support life any more than it can combustion.

The properties of hydrogen may be demonstrated by the following experiments:—

The absence of colour, taste, and odour may at once be detected by simple inspection of a jar of the gas. (It does frequently happen that the hydrogen obtained by ordinary methods, and not carefully purified, has a very disagreeable smell. This is caused by the presence of small quantities of sulphur, arsenic, or carbon, present as impurities in the zinc, which form, with hydrogen, volatile compounds of a disagreeable odour. If the zinc and sulphuric acid be perfectly pure, no odour whatever is perceived.)

Exp. 27.—Its insolubility, or rather very *slight* solubility, may be inferred from the fact that it can be collected at the pneumatic trough, without any perceptible loss of the gas; but it may be accurately measured by taking a graduated glass tube, and having put in fifty measures of water, inverting it over the mercurial pneumatic trough (fig. 13); then passing in two or three measures of hydrogen, no means whatever will make the water absorb more than one measure of the gas.

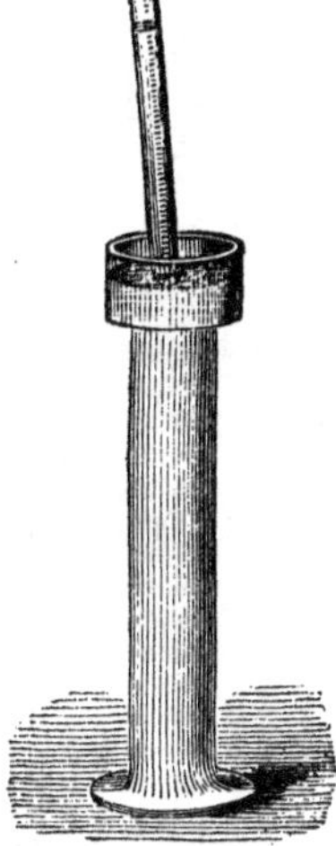

Fig. 13.

That hydrogen will not support ordinary combustion, but is itself combustible, may be shown by the following experiment:—

* Diffusion is now more generally considered as due to the intestine movements of gases, as developed in the kynetic theory of gases; for a full consideration of this, see Prof. Clarke Maxwell's *Theory of Heat*, Chap. XXIII, page 281, *et seq.*

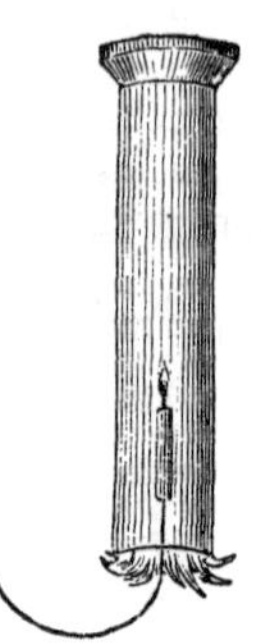

Fig. 14.

Exp. 28.—Take a jar of hydrogen, and holding it mouth downwards on account of its lightness, pass by means of a bent wire a lighted taper into it (fig. 14). The gas itself will take fire at the mouth of the jar where it is in contact with the air, but the taper will go out. (Mr. Barrett states, in the *Philosophical Magazine*, that the blue appearance of the flame is due to the presence of sulphur, either in the gas, or in the dust of the air in which it burns.)

The flame of hydrogen is a pale blue, almost invisible one; and the product, when burnt in air or oxygen gas, is water: and nothing but water. This may be shown in several ways.

Exp. 29.—If a clean, cold, dry tumbler or beaker be held over a jet of burning hydrogen, its sides will be almost immediately covered with dew, caused by the condensation of the vapour of water formed by burning *hydrogen* in the *oxygen* of the air.

Exp. 30.—It may, however, be more conclusively shown, and the water actually collected, by the following arrangement shown in fig. 15:—*a* is a tube to contain chloride of calcium to dry the

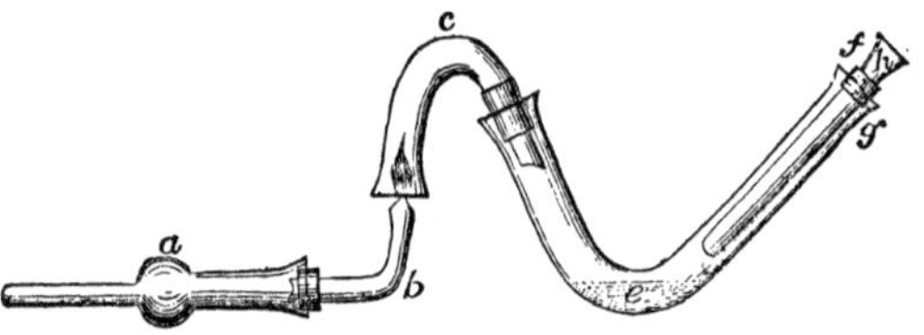

Fig. 15.

hydrogen gas; *b*, an infusible glass jet at which the gas is burnt; *c*, an infusible glass connecting tube; *e*, the condensing tube in the bend of which the water collects; *h*, a test tube filled with cold water to condense the water formed by the gas. An appreciable quantity of water will be collected at *e*, in the course of five minutes; but, if the flame (which should be about half an inch long) be kept steadily burning for about half an hour, a considerable quantity of water will be collected. The tube, *a*, should be connected (fig. 15) with the ordinary generating bottle shown in fig. 10.

In exactly the same manner water may be collected (although not in so large a quantity) from the burning of any combustible body which has hydrogen in its composition, such as alcohol, oil, coal gas, &c.

The reaction that takes place is precisely identical in all cases. The hydrogen of the burning body unites with the oxygen of the atmosphere to form water, gas, or vapour of water. This, striking on the cold sides of the glass, becomes condensed to the liquid condition.

The *lightness* of hydrogen may be demonstrated by a variety of experiments:—

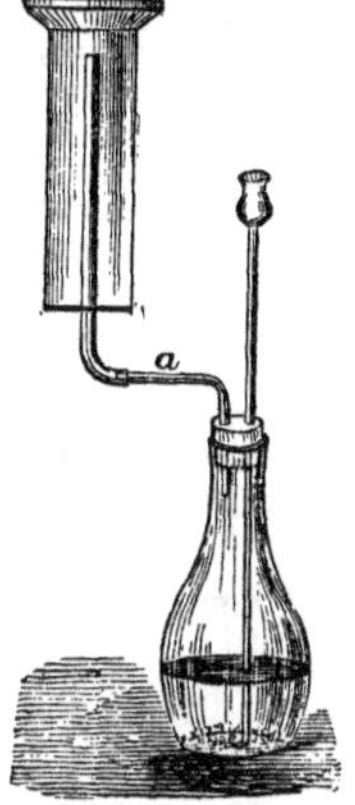

Fig. 16.

Exp. 31.—If a jar of hydrogen be collected at the pneumatic trough, it may be removed, mouth downwards, and held in that position for several minutes, without any stopper, and, on the application of a light, the gas will take fire with a very slight explosion, showing that very little hydrogen has escaped; but, if the jar be removed mouth upwards, the escape of the gas is almost instantaneous, as may be demonstrated by the application of a light.

Exp. 32.—On account of its lightness, hydrogen may be collected by upward displacement, as shown in fig. 16. The delivery tube, *a*, should pass upwards almost to the top of the jar.

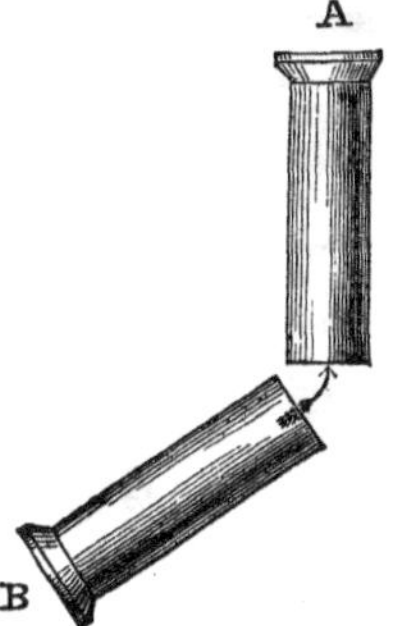

Fig. 17.

Exp. 33.—It may also, from the same cause, be decanted upwards, as shown in fig. 17. If an empty jar, A, be taken and held with its mouth downwards, and a jar of the gas, B, be carefully brought into the position shown in the figure, the hydrogen will pass from B into A, in the direction shown by the arrow, and dis-

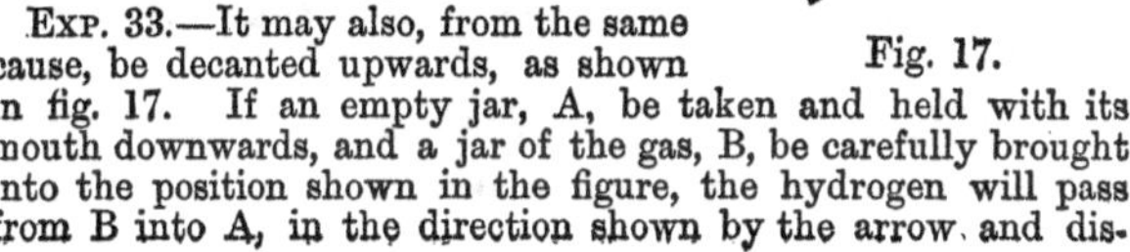

place the air in A. If the experiment has been performed neatly, on applying a light to the jar B, it will be found to be entirely free from hydrogen; but, on applying it to the jar A, a slight explosion takes place, showing that A, which only contained air, now contains a mixture of hydrogen and air, which is very explosive.

A small balloon, made of goldbeater's skin, will, if filled with dry hydrogen, ascend in the air, thus proving the lightness of the gas.

And, finally, its lightness may be easily demonstrated by actual weighing. If a large dry beaker glass be very accurately weighed in scales that will turn with a grain, then filled with hydrogen by upward displacement, and placed mouth downwards on the scales, it will be found to be sensibly and measurably lighter that it was at first when filled with air.

58. Diffusion.—This has been defined to be "the tendency of the particles of a gas to separate from each other as far as they possibly can." This power of diffusion has been found to depend on the density of the gas, being greatest in those of least density; and a series of most carefully conducted and accurate experiments, by Mr. Graham, has succeeded in establishing the following law:—

"The velocity of diffusion of different gases is inversely proportional to the square roots of their densities."

Thus, the densities of H and O being respectively 1 and 16, the square roots of which numbers are 1 and 4, their rates of diffusion will be respectively 1 and $\frac{1}{4}$—*i.e.*, H will diffuse with four times the velocity of O.

The great difference between the density of H and all other gases, makes it particularly suitable for demonstrating this property by experiment.

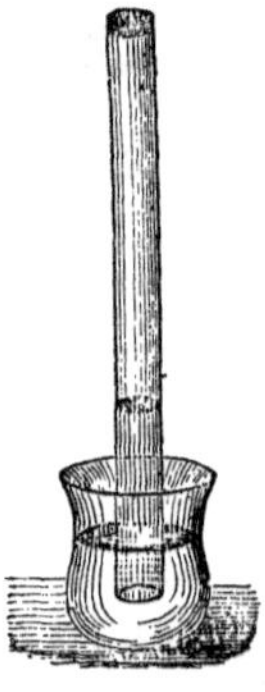

Fig. 18.

Exp. 34.—Take a glass cylinder, from 10 to 12 inches long, and from $1\frac{1}{2}$ to 2 inches in diameter; close one end with a plug of plaster of Paris, from ½ to ¾ inch thick; cover the plaster of Paris end with a plate of glass, and fill the tube with hydrogen by upward displacement; insert the lower end in a glass vessel of coloured water (fig. 18), and remove the glass plate from the top, the hydrogen will pass out so much more quickly than air can pass in, that the coloured liquid will rise in the cylinder to the height of 2 or 3 inches.

EXP. 35.—The converse of this may be shown (fig. 19) by taking a porous cylinder (such as those used for galvanic batteries), and fitting to its open end, by means of a bung, a tube about 3 feet long, and ½ inch in diameter. (Great care is necessary that the bung should be made air-tight by coating it with sealing-wax dissolved in spirits of wine.) This tube is then supported in a vertical position, so that its open extremity shall dip about an inch below the surface of some coloured water in a glass vessel. A bell-jar of hydrogen is then held over the porous cylinder, when the diffusive power of the hydrogen is made manifest by the expulsion of the air in the tube and its bubbling up through the water, caused by so much more hydrogen having entered the porous cell than the air which passed out. If the bell-jar be now removed, after all the bubbling up of the air has ceased, the hydrogen will in like manner escape from the porous cell, and the coloured water will rise in the tube to the height of 10 or 20 inches. If the experiment be repeated, substituting for the hydrogen a jar of carbonic acid or coal gas, it will be found that the heavier the gas the less will it pass through the porous cell, thus proving the truth of the general principle of diffusion.

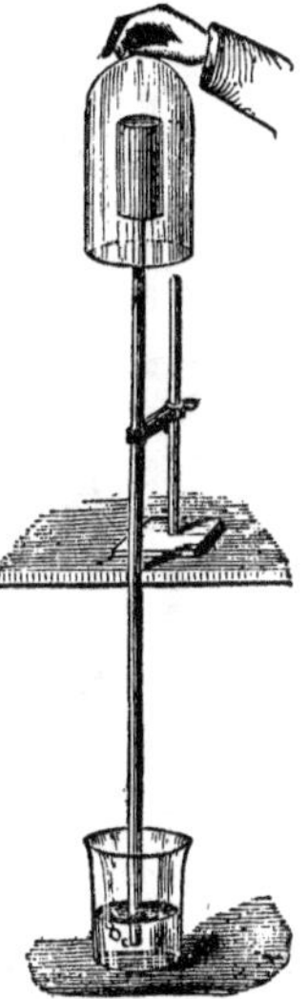

Fig. 19.

Hydrogen, in consequence of its exceeding rareness, has a peculiar effect on sound.

EXP. 36.—Fill a bell-jar with hydrogen by upward displacement, suspend it to a retort-stand, and strike a bell in it (fig. 20), scarcely any sound will be heard.

EXP. 37.—If a deep inspiration of hydrogen be taken (this may be safely done two or even three times) it will change the deepest bass voice into the thin shrill treble of a child.

59. Combinations.—The compounds which hydrogen

forms with the various elements treated of in this little work are as follows:—

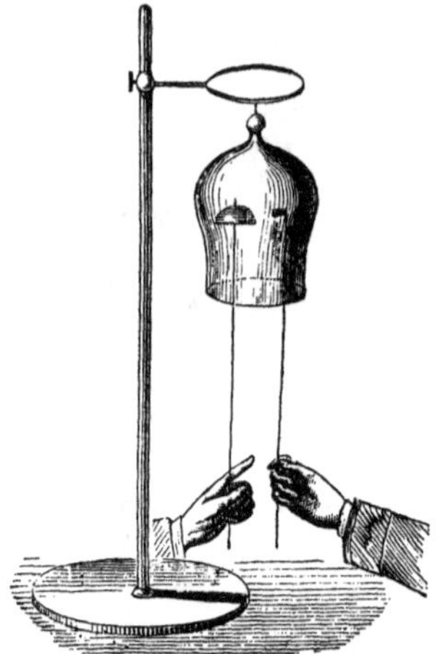

Fig. 20.

With *chlorine*, only one, *hydrochloric acid*, HCl.

With *oxygen*, two, viz., water, H_2O, and hydroxyl, H_2O_2.

With *carbon*, *hydrogen* forms many compounds called generically *hydrocarbons*. The full consideration of these belongs to organic chemistry; but two are sufficiently common and of sufficient importance to be noticed here, viz. :—

Marsh gas, or *light carburetted hydrogen*, $\mathbf{C}^{iv}H_4$.

Olefiant gas, or *heavy carburetted hydrogen*, $\mathbf{C}^{iv}{}_2H_4$.

With *nitrogen*, it forms but one compound, *ammonia*, $\mathbf{N}H_3$.

With *sulphur*, it forms two compounds—*sulphuretted hydrogen*, $\mathbf{S}H_2$, and *hydrosulphyl*, $'\mathbf{S}'_2H_2$ or Hs_2.

The full consideration of these compounds will be reserved until the elements with which the hydrogen combines shall have been respectively considered.

60. Arithmetical Considerations.—One grain of hydrogen, at the normal temperature of 32° F. or 0° C., and pressure of barometer 30 inches or 760 m.m., measures 44·4 cubic inches; and one gramme, at the same normal temperature and pressure, measures 11·2 litres.

The atomic weight of hydrogen being 1, it will require 39 grains or grammes of potassium, or 23 grains or grammes of sodium, to set free 1 grain or gramme of hydrogen. So again, zinc and iron being dyad metals, 56 grains or grammes of iron, or 65 grains or grammes of zinc, will liberate 2 grains or grammes of hydrogen

So again, with respect to the acids employed, hydrochloric and sulphuric acids being both entirely decomposed, it is a matter of simple rule of three calculation to ascertain how much must be employed to obtain any given quantity, either by weight or measure of hydrogen.

Thus, 36·5 grains or grammes of hydrochloric acid (HCl) will yield 1 grain or gramme of hydrogen; and 98 grains or grammes of sulphuric acid **S**O_2,Ho_2 [H_2SO_4] will yield 2 grains or grammes of hydrogen.

61. Hydrogen a Metal.—There are good grounds for supposing that, if hydrogen could be solidified, it would be found to be a metal,—that, in fact, the gas, as we meet with it, is but the vapour of a highly volatile metal. Thus, if the metal *palladium* be employed as the negative electrode in the voltaic arrangement for the decomposition or electrolysis of water, instead of the hydrogen being set free, it is absorbed by the palladium electrode, with which it forms a veritable alloy; and in this condition it conducts heat and electricity, and is electric, or rather, magnetic, in this respect acting exactly as it would do if it were a metal. Other metals besides palladium, such as platinum and iron, possess this property of absorbing hydrogen, but not to the same extent. This would seem to point out hydrogen as a metal, since no case of a true alloy has been known except in the case of the combination of a metal with a metal.

Fig. 21.

Another strong argument in support of this view is to be found in the fact that hydrogen gas possesses the power of conducting heat, which other gases do not. The good conductivity of hydrogen may be shown by the following experiment:—

Exp. 38.—Within a glass tube (fig. 21) is stretched a platinum wire, which is raised to incandescence by an electric current. When air, or any gas other than hydrogen, is passed through the tube from the bladder beneath, the incandescence continues; but it disappears as soon as hydrogen is employed. The heat of the wire is, in fact, conducted away by the hydrogen.

CHAPTER VIII.

Chlorine—History—Distribution and Natural History—Preparation—Properties—Combinations.

Symbol, Cl_2. Atomic weight, 35·5. Molecular weight, 71. Molecular volume, □□. 1 litre weighs 35·5 criths. Has been liquefied at 15·5 C. under a pressure of 4 atmospheres, but has never been solidified. Atomicity, ′. Evidence of atomicity, HCl.

62. **Synonymes.**—*Chlorine*, from χλωρός, green (Davy). *Dephlogisticated muriatic acid* (Scheele). *Oxymuriatic acid* (Lavoisier).

63. **History.**—Chlorine was discovered by Scheele in 1774, and from the fact of its being obtained largely from sea salt, and being associated with a something which burnt readily (hydrogen), which substance was regarded as synonymous with *phlogiston*, the chlorine gas was called *dephlogisticated muriatic acid.* Lavoisier in his system of nomenclature regarded it as a compound of *hydrochloric acid* and *oxygen*, and called it *oxymuriatic acid.* Lavoisier fell into this mistake from falsely regarding *oxygen* as the universal acidifying principle. Sir H. Davy in 1810 demonstrated that the gas did not contain oxygen, nor could it be resolved into simpler elements, and was therefore itself an element.

He applied to it the name "*chlorine*," on account of its yellowish-green tint.

64. **Distribution and Natural History.**—*Chlorine* is never found free or uncombined in nature; but in a state of combination it is a constituent of both the inorganic

ıd organic kingdoms. In the former it exists combined ith sodium, forming the large beds of salt in various .rts of the world. In the ocean it exists combined with ›dium, and also with calcium, magnesium, and potassium. n the organic kingdom it enters into the composition of ⁄arious fluids and secretions both of animals and vege- ;ables. It is also evolved from volcanoes in the form of *ıydrochloric acid.*

65. Preparation.—1. By heating certain metallic chlorides, as auric and platinic chlorides :—

$$\mathbf{Pt}Cl_4 = Pt + 2Cl_2.$$

. latinic chloride. Platinum. Chlorine.

$$[PtCl_4 = Pt + 2Cl_2.]$$

In practice this method is never adopted, *platinic chloride* being so expensive; but as an analytical experiment, or as an experiment of demonstration for a class, it is a very interesting one.

Exp. 39. Put a small quantity of *platinic chloride* in a flask or test tube having a leading tube attached, apply heat gently chlorine will be given off.

2. By *gently* heating a mixture of *manganic oxide (binoxide of manganese)* and hydrochloric acid, chlorine is given off, the reaction taking place in two stages—

(i.) $\mathbf{Mn}O_2 + 4HCl = \mathbf{Mn}Cl_4 + 2\mathbf{O}H_2.$

Manganic oxide. Hydrochloric acid. Manganic chloride. Water.

(ii.) $\mathbf{Mn}Cl_4 = \mathbf{Mn}Cl_2 + Cl_2.$

Manganic chloride. Manganous chloride. Chlorine.

In an ordinary way the reaction is given in one equation as follows :—

$$[MnO_2 + 4HCl = MnCl_2 + 2H_2O + Cl_2.]$$

Manganic dioxide. Hydrochloric acid. Manganous chloride. Water. Chlorine.

Exp. 40. Take about 3 ounces (100 grms.) of manganic dioxide (binoxide of manganese) in powder, put it in a 40-ounce flask into which you have previously well fitted a thoroughly sound good cork, through which passes a safety-funnel (which is best of the form shown in Fig. 22), and a delivery tube, bent at right angle next add half a pint, 10 oz. (about 350 c.c.) of commercial hydr-

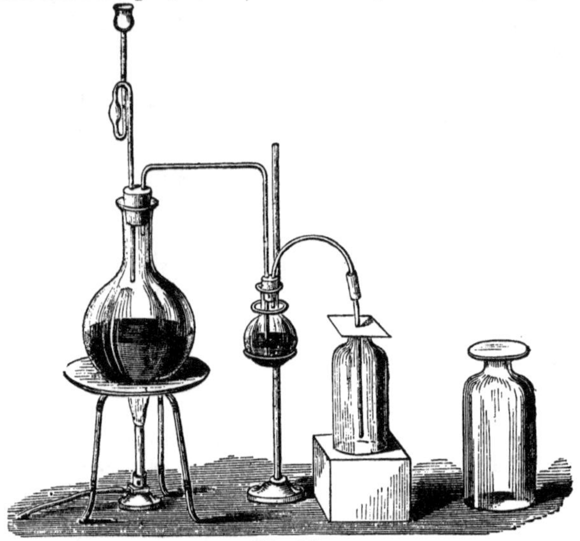

Fig. 22.

chloric acid, diluted with 3 oz. (100 c.c.) of water,* thoroughly mix the acid with the powdered manganese so as to wet the bottle throughout, then place it on a sand bath and apply heat. As chlorine does not come off quite pure, a little hydrochloric acid being given off with it, it is advisable to pass it through a wash bottle, as shown in the figure. It may then be collected at the pneumatic trough (over warm salt water), or, on account of its high specific gravity, by downward displacement. The latter course is generally preferred.

3. By heating a mixture of sulphuric acid, sodic chloride, and manganic oxide, when the whole of the chlorine present is liberated. Thus—

* This proportion of materials will give about 3 to 4 gallons of the gas, if proper care be taken to prevent its escape during collection.

$$\mathbf{Mn}O_2 + 2\mathbf{S}O_2Ho_2 + 2NaCl = \mathbf{S}O_2Nao_2.$$

Manganic oxide. Sulphuric acid. Sodic chloride. Sodic sulphate.

$$+ \mathbf{S}O_2Mno'' + 2\mathbf{O}H_2 + Cl_2.$$

Manganous sulphate. Water. Chlorine.

$$[MnO_2 + 2NaCl + 3H_2SO_4 = 2NaHSO_4$$

Manganic dioxide. Sodic chloride. Sulphuric acid. Hydro-sodic sulphate.

$$+ 2H_2O + Cl_2.$$

Water. Chlorine.]

The gas commences to come off slowly directly the mixture is made, but as soon as a very gentle heat is applied it is given off very freely. A little hydrochloric acid is formed in this reaction, but it can be easily washed out by causing the gas to pass through a wash bottle containing a little water, as shown in fig. 21, B.

4. By the electrolysis* of hydrochloric acid.

66. Properties.—At ordinary temperatures and pressures it is a gas, transparent, but of a greenish yellow colour, and obnoxious suffocating odour. (Even when largely diluted with air, if breathed, it produces distressing irritation of the air passages of the throat; in fact, causing a severe form of cold.) It is very soluble, water at a temperature of 15° C. (60° F.) dissolving twice its bulk, and even at 40° C. (104° F.) it dissolves 1·36 of its bulk.

On account of this solubility chlorine cannot be collected over cold water at the pneumatic trough, and mercury being acted on by the gas with great rapidity, that fluid cannot be substituted for the water. It is therefore generally collected over warm water, or by downward displacement in perfectly dry bottles, as shown at C, fig. 21. When this latter method is adopted it is impossible to prevent the escape of some of the gas; it is therefore advisable to carry on the operation, when possible, either in a glass closet, with an outlet, or else in the draft of an open window, in order to carry off the gas which escapes.

It is much heavier than air, 11·2 litres at 0° C., and

* For the electrolysis of hydrochloric acid see page 93.

barometer 760 m.m. pressure, weighs 35·5 grammes; (1 litre at the normal temperature and pressure weigh 3·17 grammes. Under a pressure of 4 atmospheres a 15° C. (60° F.) it has been condensed to a pale yello\ liquid, which has hitherto resisted all attempts to freeze it although exposed to a cold of $-140°$ C. ($-220°$ F.)

If a solution of chlorine (chlorine water) be exposed tc a temperature of 0° C. (32° F.) a crystalline compound of chlorine, and water will be formed (hydrate of chlorine, $Cl, 5H_2O$).

This hydrate of chlorine enabled Faraday to reduce chlorine to a liquid condition, it being one of the first gases which he so reduced. "The compressing force was the gas's own elasticity, and was thus applied. Some of the hydrate of chlorine was put into a small bent tube hermetically sealed, and a gentle heat applied. The hydrate of chlorine was by this treatment decomposed, and chlorine was liberated, which condensed into a fluid at the cold extremity of the tube, where two distinct liquids were found. The uppermost and lightest of these was merely a solution of chlorine, but the underlaying stratum was amber-coloured, and was free chlorine. . . . . . This means of effecting the condensation of gaseous bodies, namely, by the force of their own elasticity, has been successfully applied to many others besides chlorine." —Faraday's *Lectures on the Inorganic Elements*, page 121.

Chlorine is not combustible, nor does it combine directly with oxygen. The most characteristic chemical property of chlorine is its powerful attraction for many other elements at the ordinary temperature. Among the metalloids may be mentioned hydrogen, phosphorus, sulphur, arsenic, iodine, and bromine. Nearly all the metals behave in the same way especially antimony, copper, and silver.

Exp. 41. If equal quantities of hydrogen and chlorine be mixed they will, on exposure to the direct rays of the sun, or of the electric or magnesium lights, combine instantaneously with explosion; if, however, they are left for a few hours in diffused daylight, they will be found to have combined gradually.

EXP. 42. If a piece of dry *phosphorus* be placed in a deflagrating spoon and immersed in a jar of *chlorine*, the *phosphorus* will take fire, combining with the *chlorine* to form terchloride of phosphorus, PCl_3. *Note.*—Especial care must be taken to prevent the escape of any fumes, as they are very injurious.

EXP. 43. If filings of any of the following metals, in a state of fine powder, and previously heated to 30° C. (86° F.) be sprinkled in a jar of chlorine, they will take fire and burn with a rapid scintillating combustion, varying in colour according to the metal employed; thus antimony, silver, and lead give a white colour; zinc and tin, a bluish white; copper, a dull red; iron, a vivid red; arsenic, gold, and tellurium, a green. These effects are best seen if the jar in which the powdered metal is sprinkled be a tall narrow one. As the fumes resulting from these combustions are at all times inconvenient, and sometimes even dangerous, the jars should be put away as quickly as possible.

Fig. 23.

EXP. 44. The powerful affinity of chlorine for hydrogen and its indifference for carbon is strikingly shown by immersing a well lighted taper in a jar of chlorine. The taper will continue to burn, but with a dull red smoky flame. The hydrogen of the taper combines with the chlorine, but the carbon is set free as smoke, which is deposited as a fine soot.

EXP. 45. This is still more strikingly shown if a piece of blotting or filter paper be dipped in oil of turpentine, and plunged into a jar of chlorine. A brilliant flash of light is seen, and the interior of the jar is immediately covered with a thick deposit of soot.

The *rationale* of this last experiment is as follows:—*Oil of turpentine* is a hydro-carbon, having the following composition, $C_{10}H_{16}$. When, therefore, a rag or paper moistened with it is plunged into chlorine, the hydrogen and chlorine unite with evolution of light and heat, carbon being liberated. Thus—

$C_{10}H_{16}$	+	$8Cl_2$	=	$16HCl$	+	$10C$.
Turpentine.		Chlorine.		Hydrochloric acid.		Carbon.

Another important property of *chlorine* is its power of bleaching vegetable colours, and disinfecting or decomposing noxious vapours formed by organic bodies in a putrid or semi-putrid condition.

This arises from its peculiar action on all bodies containing *hydrogen.* It combines with part of the *hydrogen,* and withdraws it from the combination, each atom of *chlorine* uniting with one atom of *hydrogen,* and forming an atom of *hydrochloric acid;* whilst, at the same time, for each atom of *hydrogen* withdrawn from the original combination, an atom of chlorine is substituted. These chlorine compounds are in nearly all cases colourless and odourless. In this way *chlorine,* especially when by its combination with lime it is put into a form more easy of manipulation, has rendered incalculable service to the linen-manufacturer, the dyer, the calico-printer, and the paper-maker, and has also proved of so much use as a deodorizer or disinfectant, in all cases of contagious and infectious diseases, and of bad smells, arising from imperfect drainage.

Exp. 46.—Introduce into a bottle or cylinder, full of chlorine gas, some flowers of various colours, a strip of litmus paper, some cotton print (Turkey red), and some slips of printed and written paper. (*Note.*—All these things should be damp or moistened.) Everything but the printed paper will be affected. The flowers, the litmus paper, the cotton print, will have entirely lost their colour; the ink of the written paper will have acquired a reddish colour; while the printing-ink, being mineral, will remain unaffected. Its action on vegetable colours may be illustrated by its action on indigo. Thus—

C_8H_5NO	+	OH_2	+	$2Cl_2$	=	$C_8H_4ClNO_2$	+	$3HCl$.
Indigo.		Water.		Chlorine.		Chlorisatin.		Hydrochloric acid.

Exp. 47.—If a piece of putrid or half-putrid meat or vegetable be moistened and placed in a jar of chlorine, in a few seconds the offensive smell will have entirely passed away.

67. Combinations.—*Chlorine* forms with *hydrogen* but one combination—*hydrochloric acid* (HCl).

With *oxygen* it forms three compounds. Thus—

		Mol. Vol.	In 100 Parts. Cl.	O.
Hypochlorous anhydride,	Cl_2O = 87	[box]	81·60 +	18·40.
Chlorous anhydride,	Cl_2O_3 = 119	[box]	59·66 +	40·34.
Chloric peroxide or dioxide,	ClO_2 = 67·5	[box]	52·59 +	47·41.

With *boron* it unites in one proportion—

Boric chloride or trichloride,....BCl_3 = 117·4.

With *sulphur* it unites to form two compounds—

Sulphuric chloride, ...S_2Cl_2 = 135 [box].
Sulphuric dichloride, ...SCl_2 = 103. [box]

It is capable of uniting, directly or indirectly, with every other element with which we are acquainted.

Its use in the arts as a bleaching, and, in a sanitary point of view, as a deodorizing or disinfecting agent, has been already alluded to. In the laboratory its chief use is as an oxidizing agent. Chlorine possesses great attraction for hydrogen and all the basylous elements and radicals. It therefore decomposes water and bases, liberating *oxygen,* which in its *nascent* condition readily enters into new combinations. This is shown by the formation of *chlorisatin,* in the action of chlorine-water on *indigo,* and by the change which takes place when *chlorine-water* is exposed to direct sunlight. Thus—

$$2OH_2 + 2Cl_2 = 4HCl + O_2.$$

The decomposition of water by *chlorine* is shown by the following experiment:—

Exp. 48.—"A porcelain tube is taken, which is bound round with sheet copper to prevent it from cracking, and loosely filled with fragments of broken porcelain to expose a large heated surface. This tube is gradually heated to redness in a charcoal furnace (Fig. 24). One end of it receives the mixture of chlorine. with steam obtained by passing the chlorine evolved from hydrochloric acid and binoxide of manganese in A, through a flask, B,

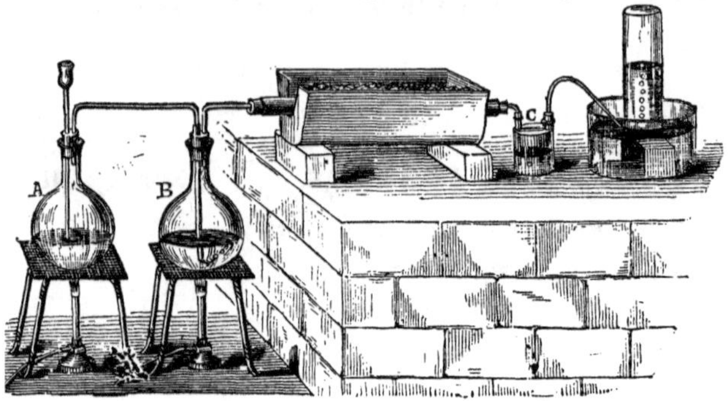

Fig. 24.

of boiling water. The other end of the tube is connected with a bottle, C, containing a solution of potash, to absorb any excess of chlorine and the hydrochloric acid formed. From this bottle the *oxygen* is collected over the pneumatic trough."—Bloxam's *Chemistry.*

CHAPTER IX.

Hydrochloric Acid—History—Distribution and Natural History—Preparation—Properties—Combinations.

Symbol, HCl. Specific gravity, 1·25. Atomic and molecular weight, 36·5. Molecular volume, $\square\square$. 1 litre weighs 18·25 criths. Has not been solidified. Under a pressure of 40 atmospheres, condenses at 10°C.

68. Synonymes.—Hydrochloric acid, muriatic acid, marine acid, chlorhydric acid, hydric or hydrogen chloride, spirit of salt.

69. History.—Discovered by Priestley in 1772. Being first obtained from sea salt, it got the name of *marine acid,* and spirits of salt.

70. Distribution and Natural History.—Found in combination in the form of muriates or hydrochlorates, and is largely evolved from volcanoes. It is also produced in enormous quantities in the alkali works; these, in fact, are the chief source of the acid for commercial purposes.

71. Preparation.—(1.) By the direct union (either by electricity or light) of equal volumes of its elements. Thus—

$$H_2 + Cl_2 = 2HCl.$$

(2.) By the action of sulphuric acid upon common salt. Thus—

$\mathbf{S}O_2Ho_2$	+	NaCl	=	$\mathbf{S}HoNao$	+	HCl.
Sulphuric acid.		Sodic chloride.		Hydric sodic sulphate.		Hydrochloric acid.
[H_2SO_4	+	NaCl	=	$NaHSO_4$	+	HCl.]

In this reaction, the sodium of the common salt changes places with the hydrogen of the sulphuric acid.

Exp. 49. Put into a gas bottle (Fig. 25), (similar to that used for hydrogen and chlorine), about 450 grs. (1 oz. troy) of common salt previously dried, and add to it through the safety funnel about 1 fluid oz. of strong sulphuric acid. On applying a gentle heat, the reaction takes place as shown in the above equation, and hydrochloric acid is given off, which, on account of its great weight, may be collected by downward displacement. The bottles in which it is collected should be *perfectly* dry, and should during the collection be covered with a piece of perforated cardboard. After collection, the jar or bottle should be covered with a *glass plate*, well smeared with grease.

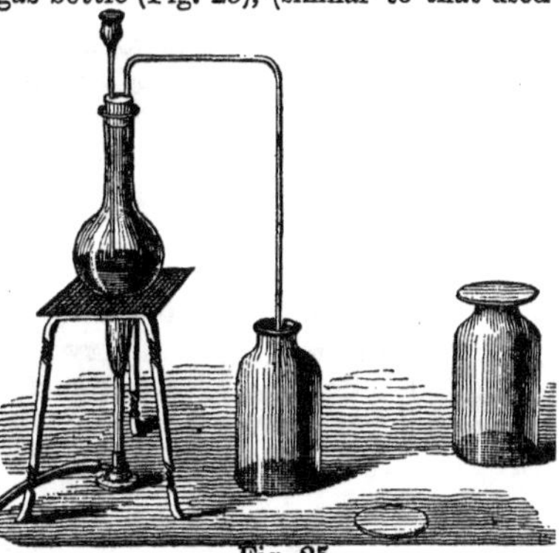

Fig. 25.

Note. 1. A stopper should on no account be used to close a jar of hydrochloric acid gas, as if there be the slightest moisture present, the gas becomes condensed and the stopper irremoveably fixed.

2. As common salt in powder frequently frothes up, to an inconvenient extent, when mixed with *sulphuric acid*, it is found in practice better to fuse the salt first, which may be done by exposing it to heat, either in a common clay crucible, or in an ordinary iron ladle.

3. It is also advisable to mix the *sulphuric acid* with half its weight of water (as great heat is generated by this mixture, allow it to cool before using) previous to pouring it on the salt. The water takes no part in the reaction, but serves to dissolve the *sodic sulphate* as it is formed, and prevent it from clogging the apparatus.

A more regular supply of hydrochloric acid gas may be obtained by gently heating about 2 oz. of *chloride of ammonium* and 2 oz. by measure of *sulphuric acid*. The change which takes place may be represented as follows:—

$$\underset{\text{Chloride of ammonium.}}{2\mathbf{N}'''H_4Cl} + \underset{\text{Sulphuric acid.}}{\mathbf{S}O_2Ho_2} = \underset{\text{Ammonic sulphate.}}{\mathbf{S}O_2(N^vH_4O)_2} + \underset{\text{Hydrochloric acid.}}{2HCl.}$$

$$[2NH_4Cl + H_2SO_4 = H_2SO_4,2H_3N + 2HCl.]$$

(3.) If moist chlorine be transmitted through a red hot porcelain tube, *hydrochloric acid* is formed and oxygen set free. Thus—

$$\underset{\text{Water.}}{2\mathbf{O}H_2} + \underset{\text{Chlorine.}}{2Cl_2} = \underset{\text{Hydrochloric acid.}}{4HCl} + O_2.$$

$$[2H_2O + 4Cl = 4HCl + O_2.]$$

72. Properties.—*Hydrochloric acid* is a transparent colourless gas, of a peculiar pungent odour, and an intensely acid taste. It has a very irritating effect on the eyes, and although not so injurious to breathe as *chlorine*, is nevertheless very much so, even when largely diluted with air. It has also a very destructive effect on vegetation.

It is heavier than air, 1 litre of the gas at 0° C. and 760 m.m. pressure weighing 1·6352 grammes. Its specific gravity is 18·25, for when equal volumes of hydrogen and chlorine combine, no alteration of volume takes place, and the specific gravity of the resulting compound is therefore the mean of the specific gravities of its constituents, thus, 1 vol. of H = 1, 1 vol. of Cl = 35·5 ; ∴ 2 vols. of HCl = 36·5, and 1 vol. of HCl = 18·25. The gas has been condensed to a colourless liquid (sp. gr. 1·27) under a pressure of 40 atmospheres, and at a temperature of 10° C. (50° F.), but this liquid has never yet been frozen.

Hydrochloric acid is not combustible, nor will it support combustion; but if a piece of potassium be heated to redness, it burns, leaving only the hydrogen, which is found to occupy only half the volume of the original gas. Thus—

$$K_2 + 2HCl = 2KCl + H_2.$$

Potassic chloride.

It may also be partly decomposed into its constituents by the electric spark. It reddens dry litmus paper, and if allowed to escape into the air, combines with the moisture of the atmosphere, producing dense white fumes. It is instantaneously absorbed by water; and if a lump of ice be placed in an atmosphere of hydrochloric acid it liquefies and absorbs the gas *very rapidly*.

This powerful attraction for water is one of the most important properties of hydrochloric acid, and may be demonstrated by a very pleasing and instructive experiment.

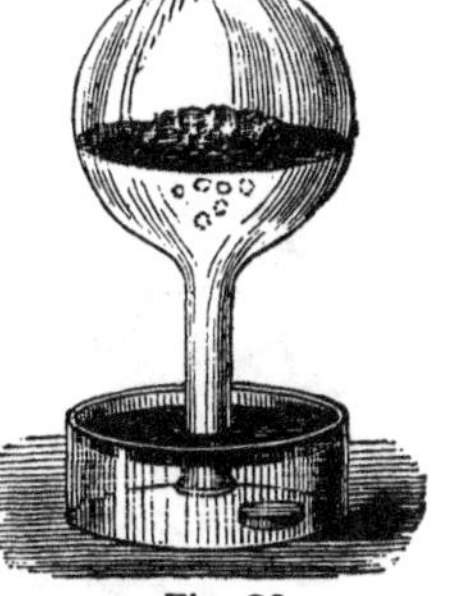

Fig. 26.

EXP. 50. The hydrochloric acid should be thoroughly dried by passing it through a wash bottle containing sulphuric acid, and can then be collected by *displacement*. A Florence oil-flask, or, better still, a globular receiver with a long neck, is used. (*Note*, This should be perfectly dry.) The delivery tube should pass down to the bottom of the flask. The hydrochloric acid, having a specific gravity as compared with the air of 1·247, gradually lifts the air out of the flask. The fulness of the flask will be noted by the escape of dense white fumes; the delivery tube should be lifted out carefully and gradually so as to prevent any air entering. When full, the flask, with its mouth downwards, should be placed in a capsule or porcelain cup of mercury. Then transfer, very cautiously, the flask, cup, and all to a glass basin of water, coloured blue by a few drops of litmus, and lift the flask out of the mercury. The water rushes in with great violence, being capable of absorbing its own weight, or about 480 times its volume of the gas, increasing in volume about one-third; and acquiring a density of 1·2109. The formation of solution of hydrochloric acid is evidenced by the instantaneous change of the blue colour to red.

73. Solution of Hydrochloric Acid.—A solution of this acid in water is one of the most useful and indispensable reagents in the laboratory, and therefore its preparation, and properties, and tests for its purity, become a matter of great importance to the chemist.

It is most conveniently prepared by the action given in the second equation on page 92. Place one part by weight of dry *sodic chloride* (common salt) previously fused, and two parts by weight of sulphuric acid (pre-

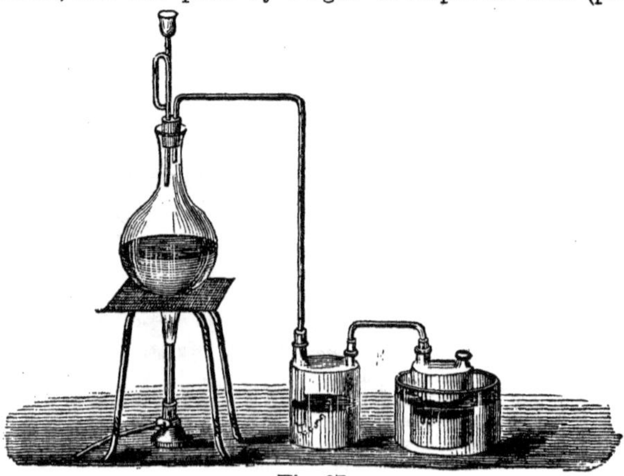

Fig. 27.

viously diluted with about one third of water) in a capacious flask or retort, shake it well together, and connect it with two (or even three) Woulfe's* bottles, and apply a gentle heat. In the first Woulfe's bottle is placed a little water, to arrest the sulphuric acid which may be carried over, as well as any other impurities which are sometimes mechanically carried over with the gas. The second bottle may contain water to dissolve the gas, and form the solution. The following precautions are, however, necessary. The bottle containing the water for the solution should be immersed in a vessel of cold water, since the absorption of hydrochloric acid by water is attended by the evolution of great heat; and, further, the tubes delivering the gas should only dip $\frac{1}{8}$ or $\frac{1}{10}$ of an inch below the surface of the water, so that the heavy solution of hydrochloric acid may fall to the bottom, and fresh water be presented to the gas.

When the solution of hydrochloric acid gas has the density of 1·21, the liquid, if pure, is fuming and colourless. If it now be subjected to an elevation of temperature, it parts with the gas freely, until the liquid which remains has a sp. gr. of 1·1 at 15·5° C. (60° F.); at this point it distils over unchanged at a temperature of 112° C. (233° F.) A weaker acid, if distilled, parts with its water freely, until it attains a sp. gr. of 1·1, when it distils over unchanged at a temperature of 112° C.

74. Impurities of the Commercial Acid, and Tests for them.—The commercial acid is liable to be contaminated by—

Iron, which gives it a yellow colour, and which may be detected by adding to it *ferrocyanide of potassium,* which gives, with the iron, a precipitate of *Prussian blue.*

Arsenic, derived from the sulphuric acid employed in its preparation. This gives, with *hydrogen sulphide,* a yellow precipitate.

Sodic chloride, which, as well as the iron, may be

* Ordinary wash bottles will do equally well.

removed by diluting the acid to a sp. gr. of 1·1 and re-distilling.

Sulphurous acid and free chlorine—these may be detected by adding, to a small portion of the acid, some *hydrogen sulphide*, when, if either or both be present, a slight turbidity will be produced.

Sulphuric acid gives, with *baric chloride*, or *baric nitrate*, a copious white precipitate.

If *pure*, the acid should leave no *residue* when evaporated. *Hydrochloric acid*, whether free or in solution, can be decomposed by many metals, hydrogen being set free. All the metals which set free water at a red heat decompose hydrogen chloride, forming a metallic chloride and setting free the hydrogen, thus—

$$Zn + 2HCl = ZnCl_2 + H_2.$$

It is in fact one of the ways in which hydrogen may be prepared.

It acts on the hydrates and oxides of most of the metals, as shown in the following equations—

(1.) **N**aHO	+	HCl	=	**N**aCl	+	**O**H_2
Sodic hydrate.		Hydrochloric acid.		Sodic chloride.		Water.
[NaHo	+	HCl	=	NaCl	+	H_2O]
(2.) **Hg**O	+	2HCl	=	**Hg**Cl_2	+	**O**H_2
Mercurous oxide.		Hydrochloric acid.		Mercuric chloride.		Water.
(3.) **Zn**O	+	2HCl	=	**Zn**Cl_2	+	**O**H_2
Zincic oxide.		Hydrochloric acid.		Zincic chloride.		Water.

—water being invariably produced whilst metallic chlorides are formed.

All metallic chlorides, with the exception of argentic, mercurous, and plumbic chlorides, are soluble in water; the latter is, however, partially soluble in cold, and readily soluble in boiling water.

"It may be useful to examine for a moment somewhat more closely into the action of *hydrochloric acid* (either in the gaseous form, or as dilute acid) upon the more important metals, and to observe—

"1. That certain metals * are readily dissolved with evolution of hydrogen, *viz.*, K′, Na′, Ba″, Sr″, Ca″, Mg″, **Fe″**, **Zn″**, Cd″, Ni″, Co″, Al[iv], and (pulverulent) Cr[iv].

"2. Others are only with difficulty soluble in boiling acid, *viz.*, Sn″.

"3. Others again are but slightly attacked by hydrochloric acid, *viz.*, Ag′, Pb″, Cu″, Sb‴, Bi‴.

"4. A few metals are not affected by either hot or cold hydrochloric acid, *viz.*, Au, Pt, As, and Hg, and (crystalline) Cr."—Valentine's *Practical Chemistry.*

75. Analysis of Hydrochloric Acid.—

EXP. 51. Hydrochloric acid may be analyzed by means of a current of electricity. If the platinum, or, better still, the carbon electrodes of a battery, be plunged in a vessel containing HCl (Fig. 28), gas is immediately given off. Its smell and bleaching power demonstrate the presence of *chlorine*, and its inflammability gives suspicion of the presence of hydrogen. By conducting the process in a V-tube (Fig. 29) closed at one end, or in any of the numerous forms of voltameters, each gas may be collected separately, and it will be found that the gas given off at the *positive pole* is chlorine, and that at the negative pole *hydrogen.*

Fig. 28.

* Practically no other but the few metals printed in thick type would ever be thus used for the preparation of metallic chlorides. The atomicity marks placed on the right hand side above the symbols indicate the nature of the chlorides which the different metals form under the given circumstances.

76. Reactions of Hydrochloric Acid.—1. Hydrochloric acid may be readily recognized by its producing, with *nitrate of silver*, a white precipitate of chloride of silver

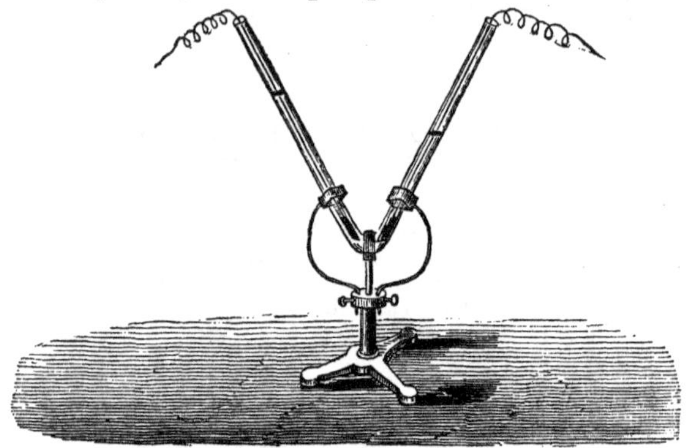

Fig. 29.

($AgCl$), which on exposure to light turns first violet and then black. It ($AgCl$) is insoluble in nitric acid, but soluble in ammonia and potassic cyanide.

2. HCl produces, with either acetate or nitrate of lead, a white precipitate of chloride of lead ($PbCl_2$), soluble in an excess of boiling water.

77. Uses.—Hydrochloric acid is largely employed in the manufacture of chloride of lime (bleaching powder), and also in the preparation of potassic chlorate and sal-ammoniac (chloride of ammonium). It is also employed by the dyer and calico printer as a solvent for tin, which is largely used by both.

CHAPTER X.

Oxygen—History—Distribution and Natural History—Preparation—Properties—Allotropic Oxygen or Ozone.

78. Oxygen.—Symbol, O. Atomic weight, 16. Molecular weight, 32. Molecular volume, [ ] . 1 litre weighs 16 criths. Atomicity, ″. Evidence of atomicity—

Water,	$\mathbf{O}H_2$ or $[H_2O.]$
Potassic hydrate,	$\mathbf{O}KH$ or $[KHO.]$
Argentic oxide,	$\mathbf{O}Ag_2$ or $[Hg_2O.]$
Hypochlorous anhydride,	$\mathbf{O}Cl_2$ or $[Cl_2O.]$

79. Synonymes.—Oxygen (ὀξύς, acid; γέννάω, I beget or give rise to), (Lavoisier); Dephlogisticated air (Priestley); Empyreal air (Scheele). Vital air.

80. History.—Oxygen gas was discovered by Priestley, August 1, 1774, and called by him *dephlogisticated air;* it was also discovered soon afterwards by Scheele in Sweden, who, however, had no knowledge of its previous discovery by Priestley. Scheele gave to it the name of empyreal air. Lavoisier, in giving it the name *oxygen,* did so under the idea that it was the acidifying principle in all acids; and although we now know this idea to be a false one, we retain the name, as it is both sufficiently distinctive and expressive.

81. Distribution and Natural History.—Of all known substances oxygen is the most widely distributed throughout nature, constituting at least three-fourths of the terraqueous globe. It occurs in the free state in the atmosphere, of which it constitutes 20 per cent. ($\frac{1}{5}$) by volume, or 23 per cent. by weight. It occurs, combined with *hydrogen,* in water, forming eight-ninths of that fluid. It is found in most mineral bodies, probably forming one-third by weight of the solid crust of our globe, for silica, carbonate of lime, and alumina—the three most abundant constituents of the earth's strata—contain each about half their weight of oxygen; and, finally, it is found in almost all animal and vegetable compounds, being an essential constituent of all living beings.

82. Preparation.—Most frequently prepared by applying heat to some substance which contains it, and which will part with it with sufficient ease.

1. If mercuric oxide, $Hg''O$, be strongly heated, it is resolved into its elements, thus—

$$\underset{\text{Mercuric oxide.}}{2Hg''O} = \underset{\text{Mercury.}}{2Hg} + \underset{\text{Oxygen.}}{O_2}.$$

Exp. 52. Put into a *hard* German glass retort some oxide of mercury,* HgO, apply heat, and the oxygen gas will be given off, and may be collected at the pneumatic trough, as shown in Fig. 31, while metallic mercury will be condensed on the cold parts of the retort.

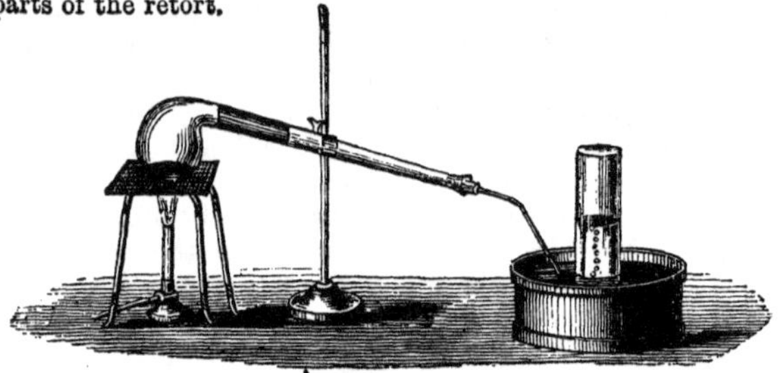

Fig. 31.

This is a very pretty class experiment, demonstrating the analysis of a compound into its elements by the application of heat, and is also of interest as the means whereby both Priestly and Scheele first isolated oxygen gas; but it is much too expensive and inconvenient for ordinary use, or for obtaining the gas in other than in very small quantities. HgO has the atomic weight 216; thus, $Hg=200$, $O=16$, $\therefore HgO=216$.

Therefore for every 16 grains or grammes of oxygen we wish to liberate, we must use 216 grains or grammes of HgO.

2. By heating native manganic oxide (pyrolusite) a portion of its oxygen is liberated. Thus—

$$\underset{\text{Manganic oxide.}}{3\mathbf{Mn}O_2} = \underset{\text{Trimanganic tetroxide or Manganoso-manganic dioxide.}}{{}^{iv}(\mathbf{Mn}_3)^{viii}O_4\dagger} + \underset{\text{Oxygen.}}{O_2}.$$

$$[3MnO_2 = Mn_3O_4 + O_2.]$$

* HgO may be made by heating metallic mercury to its boiling point with excess of air, oxygen is gradually absorbed, and a red powder (mercuric oxide, Hg''_2O) is formed.

† Trimanganic tetroxide $[{}^{iv}(\mathbf{Mn}_3)^{viii}O_4]$ will be better understood from its graphic formula, 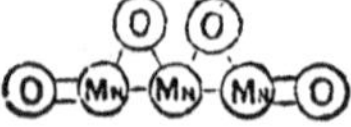

This method is adopted when oxygen is required in large quantities and expense is a matter of consideration; but as there are many difficulties in the way, it is not generally made use of at the lecture table, and but seldom in the laboratory.

The *manganic oxide* should be exposed to a red heat, when it parts with one-third of its oxygen, a reddish brown oxide of manganese remaining behind, according to the above equation. For this purpose a gun barrel, stopped at one end, and with a leading pipe attached to the other, to convey the gas to the pneumatic trough; or, better still, an iron bottle, such as is used for the importation of mercury into this country, is employed, having an iron pipe fitted into the neck by grinding, and a flexible tube attached to the other end of this pipe, to convey the gas wherever it is required. The bottle can then be heated in a furnace. The oxide of manganese usually contains some water, as well as calcic carbonate and calcic nitrate, so that the gas which comes off, especially at first, is largely adulterated with steam, carbonic anhydride, and nitrogen; the steam and carbonic anhydride can be eliminated, but it is not so easy to deal thus with the nitrogen, which contaminates the resulting gas to the extent of 5 or 6 per cent.

Black oxide of manganese furnishes, when pure, about one-ninth of its weight of oxygen; but the commercial oxide seldom yields more than one-eighteenth or one-twentieth of its weight.

3. Very pure oxygen gas may be obtained by the action of heat on potassic chlorate. Thus—

$[O_3KCl$	=	KCl	+	O_3.
Potassic chlorate.		Potassic chloride.		Oxygen.
$[KClO_3$	=	KCl	+	$O_3.]$

This method requires great heat, greater than glass vessels will generally bear, and is therefore seldom adopted in practice.

4. It is found, however, that if the potassic chlorate be

mixed with from $\frac{1}{4}$ to $\frac{1}{2}$ of its weight of some other of the metallic oxides, such as oxide of copper, oxide of manganese, &c., and even powdered glass and sand, the oxygen is given off at a much lower temperature, although such oxides have not been proved to experience any chemical change during the operation.

The equation to this will be the same as the last—the manganic, or other oxide taking no part in the reaction.

5. By the electrolysis of water.

This method—which has been explained in Chapter VII., page 73, and will be further alluded to in Chapter XI. on water—is had recourse to when it is intended to demonstrate the composition of water by analysis, or to obtain the oxygen absolutely pure.

6. By the action of heat on a mixture of manganic oxide and sulphuric acid—

$$\underset{\text{Manganic oxide.}}{\mathbf{Mn}O_2} + \underset{\text{Sulphuric acid.}}{\mathbf{S}O_2Ho_2} = \underset{\text{Manganous sulphate.}}{\mathbf{S}O_2Mno''} + \underset{\text{Water.}}{\mathbf{O}H_2} + O.$$

$$[MnO_2 + H_2SO_4 = MnSO_3 + H_2O + O.]$$

7. By passing steam and chlorine through a red hot porcelain tube, hydrochloric acid and oxygen are formed—

$$2\mathbf{O}H_2 + 2Cl_2 = 4HCl + O_2.$$

See Chapter VIII, page 89.

Oxygen may be obtained from a number of other substances, but those mentioned are the best, and are the materials and methods most usually employed, especially the methods 2, 3, and 4. Red lead and the peroxides of most of the metals when *strongly heated* furnish the gas, as do also the nitrates of potassium, sodium, and barium.

Boussingault proposed to obtain oxygen from the air by the action of *baryta*, BaO, on the atmosphere. BaO when heated to dull red heat absorbs *oxygen* from the air, becoming converted into barium dioxide, BaO_2, which at a still higher temperature yields up its second proportion of oxygen, and is reconverted into baryta, BaO. There are difficulties however attending the application of this principle, and it has hitherto not been practically successful.

83. Properties.—Oxygen is a gas, colourless, tasteless, inodorous, and permanently elastic (that is, it has never yet been liquefied). Of all known substances it exerts the smallest refracting power on the rays of light. It possesses decided, though weak, magnetic properties, like those of iron; and, like that metal, its power of magnetization can be diminished or even suspended by a very slight elevation of temperature. When tried by litmus or turmeric paper, it shows neither acid nor alkaline reaction; it does not precipitate lime water; it is very slightly soluble in water—at 62° F., 100 parts of water dissolve 3 parts of oxygen. It does not burn itself (except in an atmosphere of hydrogen), but is the great supporter of ordinary combustion and of life. It is heavier than the atmosphere, having, according to Regnault, a sp. gr. of 1·10563—1 litre at 0° C., and 760 m.m. pressure weighing 1·4298 grms., or 100 cubic inches at 62° F., and 30 inches barometer, weighing 34·203 grains.

The chief chemical character of oxygen is its intense power of combining with other substances, and, consequently, its importance in supporting combustion, and its absolute necessity in maintaining life.

84. Tests for Oxygen Gas.—*Nitric oxide*, N_2O_2, in presence of free *oxygen*, combines with it eagerly, forming dense red fumes of *nitrous anhydride*, N_2O_3, which rapidly dissolve in any water or moisture which may be present, forming a solution of *nitrous acid*, $2NHO_2$.

A solution of *potassic hydrate*, KHO, in *pyrogallic acid*, rapidly absorbs oxygen, turning black. These two reactions, therefore, serve as infallible tests for the presence of O.

Exp. 53. Fill a small graduated tube, three parts full of oxygen, at the mercurial pneumatic trough, C, Fig. 33, then pass into it, either by means of the curved pipette A, or the syringe B, a solution of *pyrogallic acid* in water, and next by the same means some strong solution of *potassic hydrate* (caustic potash). An immediate blackening of the solution will take place, the gas will be absorbed, and the mercury will rise up and take its place.

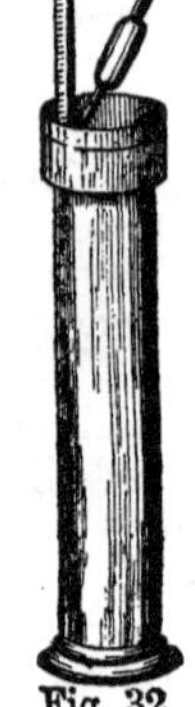

Fig. 32.

The intense powers of combination which oxygen possesses, and the products formed by its union with the other elements, may be illustrated by a series of brilliant and instructive experiments.

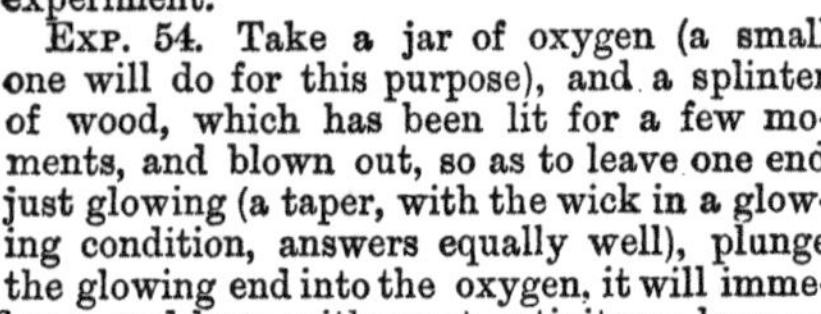

Collect six or seven jars of oxygen at the pneumatic trough. Some of them should be rather large jars, and one is better if it be in the globular form, as will be explained in the experiment.

Exp. 54. Take a jar of oxygen (a small one will do for this purpose), and a splinter of wood, which has been lit for a few moments, and blown out, so as to leave one end just glowing (a taper, with the wick in a glowing condition, answers equally well), plunge the glowing end into the oxygen, it will immediately burst into flame, and burn with great activity as long as there is any oxygen left in the jar. This is very characteristic of *oxygen*, only one other gas, *nitrous* oxide, acting in the same way. Test now the contents of the jar with moistened blue litmus paper, and it will be found to redden the paper, proving the presence of an acid; next, pour in some lime water, which will immediately turn milky, proving the acid to be *carbonic anhydride*, (CO_2), formed by the union of the *oxygen* with the carbon of the wood or taper.

Exp. 55. Take a piece of wood charcoal (if it be splintery, a better effect is produced) and having thoroughly ignited it at a lamp, plunge it into a jar of the gas (this time a larger jar should be taken), when a most brilliant combustion will take place, and continue until all the charcoal or all the oxygen is consumed. Test the product as before with litmus paper and lime water, and it will be found to be *carbonic acid.*

Exp. 56. If a piece of sulphur, about the size of a nut, be heated in a deflagrating spoon, *a*, fig. 34, and when well alight, plunged into a jar of oxygen, standing in a tray of water, it burns with a pale violet flame of great beauty. *Slight* white fumes are formed during the burning, but these are rapidly dissolved in the water in the tray. If, now, the contents of the jar, or the water in the tray, be tested with blue litmus paper, it will be found to be strongly acid in its character.

Exp. 57. The burning of phosphorus in oxygen is perhaps one of the most brilliant experiments of combustion. For this purpose a glass globe should be taken, and filled with oxygen, the *phosphorus*, a piece about the size of a large pea, should be placed on an iron cup, standing in a tray of water about 2 inches deep,

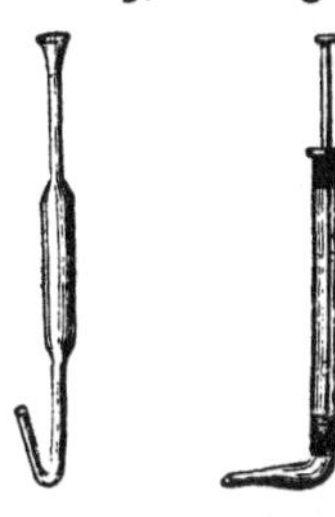

Fig. 33.

Fig. 34.

and kindled by being touched with a hot wire, the globe of oxygen should be placed quickly over it, when the most brilliant combustion will take place; dense white fumes are formed, which become so intensely luminous that the eye can scarcely gaze on them. After a little time these fumes will condense in the water in the tray, which on examination will be found to be intensely acid in its reaction.

Note.—Phosphorus should never be handled with the naked fingers, should always be cut under water, and should be always dried between blotting or filter paper before a light is applied. Neglect of these precautions may result in a serious accident.

All the above bodies are non-metals, and are also known as being in the ordinary sense of the word, "combustible bodies." It will be noted that they all form *acids*. Oxygen unites with all the non-metals except *fluorine*, and forms with them, with the exception of fluorine and hydrogen, either acids or bodies which in their reactions behave like acids.

But oxygen also unites other elements not usually considered combustible, as for instance, zinc, iron, copper, and many other metals.

Exp. 58.—If a piece of zinc foil be made into the form of a tassel, the ends gently warmed, dipped into flowers of sulphur, or sulphur in powder, then ignited, and plunged into a jar of oxygen gas, the zinc will be seen to burn with great brilliancy, giving rise to a beautiful greenish flame. On withdrawing the remainder of the tassel it will be found to have been converted into a friable mass, yellow while hot and white when cold—*oxide of zinc* (ZnO).

On testing with the test papers it will be found not to possess either the properties of an acid or an alkali; it belongs, in fact, to that large class of bodies called bases, which although not soluble in water like the alkalies, are, nevertheless, capable of combining with the acids, and neutralizing them either partially or entirely.

Fig. 35.

EXP. 59.—If a piece of iron rod, as thick as a lead pencil, be heated to redness in a furnace, and then held in front of a jet of oxygen, it will burn with great brilliancy, throwing off sparks and dropping melted oxide of iron.

The experiment is, however, generally performed by twisting ten or twelve strands of fine iron wire (pianoforte wire) into a rope, and coiling it up into a spiral, or by softening a piece of watch spring in a flame until it has lost its elasticity, and coiling it up into a spiral; one end of the spiral is then fixed into a cork which fits the neck of the deflagrating jar, and the other end is filed clean, heated, and tipped with sulphur. It is then lit, and inserted in a jar of oxygen. The burning sulphur supplies the heat to raise the iron to the point of combustion, and the wire takes fire, burning with the most brilliant coruscations, and dropping molten drops of oxide. So great is the heat generated that these drops generally, in spite of the cooling effect of the water through which they pass, melt themselves into the dish in which the deflagrating jar stands. On testing both the contents of the jar and the tray with test papers, they are found to be neutral to both acid and alkaline reactions.

Fig. 36.

Thus, we have oxygen entering into combination with most of the elements * (it enters into *direct* combination with all but Cl, Br, I, F, Au, Ag, Pt, and some few others of no practical importance, forming with some *acids*, with others *neutral bodies*, and with a third division *alkalies*).

All the *non-metals*, except *hydrogen* and *fluorine*, form

* At present no combination has been effected between *oxygen* and *fluorine*, but we have every reason to suppose that if *fluorine* could be isolated *oxygen* would be found to combine with it.

with *oxygen* one or more compounds, which, when soluble in water, possess acid properties. Many of the *metals*, however, by their union with oxygen, give rise to bodies of an opposite kind called **bases.** These basic oxides are not all of them soluble in water, but they all of them possess the property of neutralizing an acid. Intermediate between these basic and acid oxides is a third class, which does not readily enter into combination with either acids or bases, and are called **saline oxides** from their analogy in composition to salts. Of these we may instance the *black oxide of manganese*, MnO_2 or MnO,MnO_3, the *magnetic oxide of iron*, Fe_3O_4 or FeO,Fe_2O_3, and *red lead*, $2PbO,PbO_2$.

At a temperature of 0° C. (32° F.), and pressure of the barometer equal to 760 m.m. (30 inches), 11·2 litres of oxygen weigh exactly 16 grammes, or 44·4 cubic inches of the gas weigh 16 grains. At a temperature of 15·5° C. (60° F.) 46·84 cubic inches weigh 16 grains.

Knowing therefore the relation between volume and weight under certain given conditions, it is a simple rule of three calculation, to ascertain either the volume of a given weight, or the weight of a given volume, under like conditions.

If the conditions of temperature and pressure vary, calculate the required volume or required weight under the normal condition, and then correct it (1.) for the given temperature, (2.) for the given pressure, according to the examples given in Chapter III., page 36.

(1.) What will be the volume of 100 grains of oxygen, at the normal temperature (0° C.) and pressure 760 m.m.?

16 grains of O at 0° C., and 760 m.m. pressure occupy 44·4 cub. in.

∴ 1 ,, ,, ,, $\frac{44 \cdot 4}{16}$ cub. in.

and 100 ,, ,, $\frac{\overset{11 \cdot 1}{44 \cdot 4}}{\underset{4}{16}} \times \overset{25}{100} = 277 \cdot 5$ cub. in.

(2.) Required the weight of 100 litres of oxygen gas at the normal temperature and pressure—

11·2 litres of O weigh 16 grammes.

$\therefore$ 1 litre of O weighs $\frac{16}{11 \cdot 2}$ grammes.

and 100 litres of O weigh $\frac{16}{11 \cdot 2} \times 100 = \frac{16}{112} \times 1000 = 142\frac{6}{7}$ grammes.

(3.) 50 litres of oxygen were measured off at 20° C. What would be the volume of the gas at 0° C., the pressure remaining unchanged?

One litre of oxygen measured off at 0° C. becomes

$$1 + 20 \times \cdot 003665 = 1 \cdot 0733, \text{ at } 20° \text{ C.},$$

and as the contraction on cooling is equal to the expansion on heating, we have the proportion—

Bulk at 20° of 1 litre of gas measured off at 0°.		One litre.		Measure of given oxygen at 20°.		Measure required at 0°.
1·0733	:	1	::	50	:	x.

$$\therefore x = \frac{50}{1 \cdot 0733} = 46 \cdot 585 \text{ litres.}$$

So again, in the calculation of the quantity of any salt required to be decomposed to obtain a given quantity (by weight or volume) of the gas, or *vice versa*, the quantity of gas, either by weight or volume, which may be obtained from a given quantity of the salt. Find the atomic weight of the salt, and notice from the equations the amount of oxygen which that gives off, and you have the means of constructing your proportion. Thus—

$$HgO = Hg(200) + O(16) = 216, \text{ which yields 16 of O.}$$

$\therefore$ 216 grains or grammes of HgO yield 16 grains, or 44·4 cubic inches of O, or 16 grammes, or 11·2 litres of O.

So $KClO_3$ = 122·5; thus, K = 39, Cl = 35·5, O_3 = 48, and yield, when entirely decomposed, 48 of O.

$\therefore$ 122·5 grains of $KClO_3$ yield 48 grains or 133·2 cub. in. of O, and 122·5 grammes of $KClO_3$ yield 48 grammes or 33·6 litres of O, and so on.

(4.) What weight and what volume of oxygen can be obtained from 75 grains of potassic chlorate, ($KClO_3$)?

Grains.		Grains.		Grains of O.		
122·5	:	75	::	48	:	x.

$$\therefore x = \frac{75 \times 48}{122 \cdot 5} = \frac{144}{4 \cdot 9} = 29 \cdot 38 \text{ or } 29\tfrac{1}{2} \text{ grains nearly.}$$

Grains.		Grains.		Cubic inches.		
122·5	:	75	::	133·2	:	x.

$$\therefore x = \frac{75 \times 133 \cdot 2}{122 \cdot 5} = \frac{399 \cdot 6}{4 \cdot 9} = 81 \cdot 551 \text{ cubic inches.}$$

85. Ozone.—Symbol, O_3. Molecular weight, 48. Molecular volume, [ ] . 1 litre weighs 24 criths.

86. History.—Whenever an electric machine is worked a peculiar smell will be found to pervade the apartment, which for a long time was known as the electric smell. Professor Schönbein of Bâle demonstrated that this smell and its corresponding agency existed independently of its electric source, that it could be generated at pleasure, and by various means, and gave to it the name of *Ozone* (from ὄζω, I smell.)

It is now regarded as an allotropic condition (from ἄλλος, another; and τρόπος, form), of *oxygen, i. e.*, that *oxygen* is in some way modified, or made to assume a more active condition; thus, in ordinary *oxygen*, the molecule consists of two atoms, and is represented by

weighing 32;

whereas in allotropic *oxygen* or *ozone* the molecule consists of three atoms, and is represented by

weighing 48;

these three atoms or volumes being condensed into two, so that its atomic or volume weight is equal to 24.

87. Preparation.—(1.) By transmitting electric sparks through air or oxygen. A current of oxygen is passed through a tube, in which are sealed a pair of platinum wires, with the points a little distance apart. When one wire is in connection with the prime conductor of an electrical machine, and the other with the earth, as long as the machine is in action the odour of ozone is perceptible in the issuing gas, although but a very small proportion of the oxygen is acted on.

(2.) If a stick (about 1 inch long) of phosphorus be scraped so as to make it clean, and placed in a wide-mouthed bottle with a little water at a temperature of from 15° to 21° C. (60° to 70 F.), the slow oxidation of the phosphorus is attended by the production of *ozone*, which in the course of two hours attains its maximum.

(3.) By passing an electric current through *dilute sulphuric* or *chromic acid.*

Ozone, when prepared in either of these ways, is always largely mixed with air or oxygen.

Siemens prepares ozone by induction. A kind of

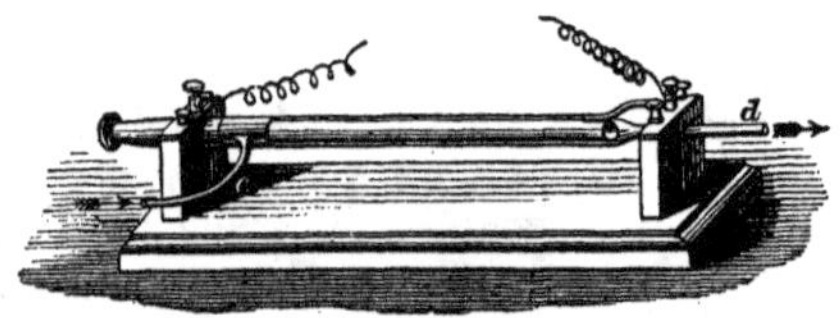

Fig. 37.

Leyden jar is made by coating the inside of a long glass tube with tin foil; over this tube is passed a second wider tube, also coated with tin foil, but on its outer surface. A current of pure dry oxygen gas is passed between the tubes; this gas becomes electrified by induction, connecting the inner and outer coating with the terminal wires of an induction coil.

88. Properties.—Ozone is a gas invisible and colourless, but with a faint sickly odour. It is insoluble in water, and in solutions of either acids or alkalies, but is absorbed by a solution of *potassic iodide.* It possesses powerfully oxidizing properties, and, consequently, even when largely diluted with air, exercises an irritating effect on the respiratory organs, and at ordinary temperatures oxidizes and destroys organic matter. *Ozone* possesses considerable bleaching powers, and converts blue indigo into *isatin.* It rapidly oxidizes silver, iron, and copper when moist. *Ozone* displaces iodine from its combination with the metals, setting the iodine at liberty; indeed, this reaction is so easily produced that it furnishes the most ready means of detecting the presence of *ozone.*

89. Tests.—Test or filter paper, soaked in a solution made of 1 part of *potassic iodide* in 200 parts of distilled water, and thickened with 10 parts of white starch, forms an exceedingly delicate test for the presence of *ozone.* The ozone liberates the iodine, which immediately combines with the starch, and forms the blue iodide of starch so characteristic of iodine.

Paper also, soaked in a solution of *manganous sulphate,* $MnSO_4$, shows the presence of ozone by becoming brown, owing to the manganese in the sulphate absorbing oxygen, and becoming converted into the insoluble hydrated peroxide, sulphuric acid being set free. If paper stained black with *plumbic sulphide,* PbS, be exposed to *ozone,* the stain will gradually disappear, the lead and the sulphur will both absorb oxygen, and white *plumbic sulphate,* $PbSO_4$, will be formed.

Ozone has its peculiar properties slowly destroyed at a temperature not much exceeding that of boiling water, 100° C. (212° F.), while at 300° C. (572° F.), the change is instantaneous.

When ozone is absorbed by a metal, no contraction of volume takes place, which can only be explained on the ground that ozone consists of three volumes of oxygen condensed to two, that the metal combines with one volume, and forms an oxide, while the other two are set free as oxygen.

Some experiments of Soret favour this supposition, but at present they cannot be considered quite conclusive.

CHAPTER XI.

COMBINATIONS OF OXYGEN.

Formation and Reactions of Water—Preparation and Properties of Hydroxyl—Compounds of Chlorine with Oxygen and Hydroxyl.

90. Water.—Symbol $\mathbf{O}H_2$ (H)—(O)—(H) [H_2O]. Molecular weight = 18. Molecular volume, [ ]. Relative weight = 9. 1 litre of water vapour, weighs 9 criths. Sp. gr. as vapour, 0·622; as liquid, 1·000; as ice, 0·918. Fuses at 0° C.; boils at 100° C.

91. History.—The ancients considered water to be an element, one of the four elements of which they believed the earth to be composed. Towards the close of the last century it was found that, when hydrogen was burned in air, water was formed; and a little later, Cavendish and Watt each demonstrated the real composition of water.

92. Occurrence.—Most abundantly throughout all nature, both organic and inorganic.

93. Preparation.—By the direct union of O and H. Whenever H is burnt in air or O, water is formed, as shown in Chapter VII., page 66.

If dry H be passed over a metallic oxide heated to redness, the H takes O from the oxide, forming H_2O, and leaving the metal. Cupric oxide is well suited for this purpose. The change that then takes place is represented by the following equation:—

$$CuO + H_2 = \mathbf{O}H_2 + Cu.$$

Exp. 60.—This method of reducing *cupric oxide* may be made use of to demonstrate synthetically the composition of water. Thus, in Fig. 38, A is a flask for the generation of hydrogen; B is a wash-bottle, containing sulphuric acid, through which the gas is made to pass in order to dry it; C is a tube, containing *calcic chloride*, to retain any traces of moisture which may have come off from the wash-bottle, B; D is a combustion tube, containing oxide of copper exposed to heat from the Bunsen burner; E, a U-tube, containing chloride of calcium, to absorb the water formed; F, a tube to carry off any excess of gas.

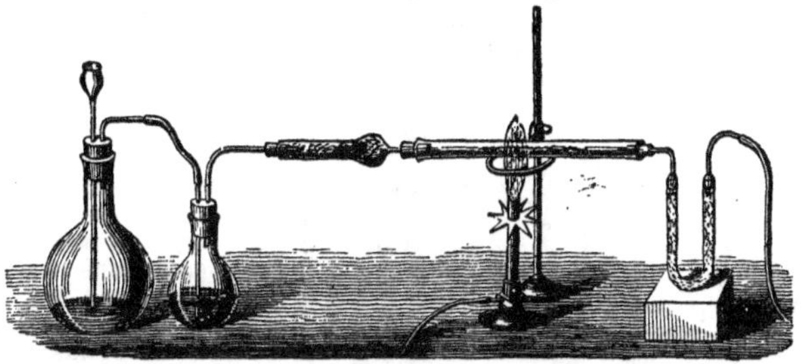

Fig. 38.

The whole of the apparatus, D, E, and F, with the *cupric oxide* and the *calcic chloride*, are carefully weighed before the commencement of the experiment, and their weight noted—D by itself, and E and F by themselves. The *cupric oxide* in the bulb of D is brought to a state of redness, and at the same time *hydrogen* is disengaged from the flask, A, by the action of dilute sulphuric acid on zinc. The *hydrogen*, in passing over the heated *oxide of copper*, absorbs *oxygen*, and forms water in the state of vapour, which becomes condensed in the U-tube E, and is absorbed by the calcic chloride. After about ten or fifteen minutes the operation may be stopped, the tubes detached and weighed. The tube D will be found to have lost weight (the amount of the oxygen abstracted from the *oxide of copper*); while the tubes E and F will be found to have gained weight (the amount of water formed and intercepted by the *calcic chloride*); and the amount of loss by the *cupric oxide* will be (in all cases) to the amount of gain by *calcic chloride*, as 16 is to 18—that is to say, that water is composed of 16 of *oxygen* to 2 of *hydrogen*.

2. Water also occurs as a secondary product in numerous chemical reactions, as, for instance, in the action of hydrochloric acid on potassic hydrate:—

$\mathbf{O}KH$	+	HCl	=	$\mathbf{O}H_2$	+	KCl.
Potassic hydrate.		Hydrochloric acid.		Water.		Potassic chloride.

94. Composition of Water.—Its composition by weight has been synthetically shown by Experiment 60, in which 18 parts by weight of water were proved to have been formed by 16 parts by weight of *hydrogen*, and 2 parts of *oxygen*.

The *quantitative* determination of the volumetric composition of water is usually conducted as follows:—

Exp. 61.—A long straight tube of thick glass, very accurately graduated, and closed at one end, through which two platinum wires have been melted, called a *eudiometer* (εὔδιος, *clear*, and μετρον, *a measure*) is taken, filled with mercury, and the open end plunged beneath a mercurial pneumatic trough. A quantity of hydrogen gas is passed into it, and its volume noted, and then a quantity of oxygen is passed in, whose volume is also noted. The gases are now exploded by means of the electric spark passed through the platinum wires; condensation immediately takes place, the sides of the tube become covered with moisture, and the mercury rises to take the place of the exploded gas.

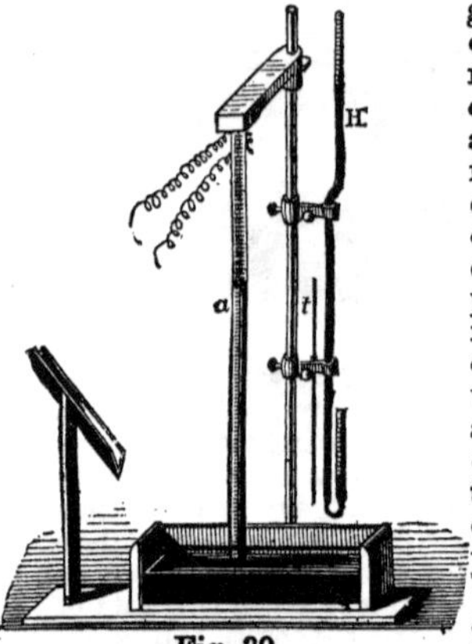

Fig. 39.

If the gases have been mixed otherwise than in the proportion of 2 of H to 1 of O, a residuum is left, and this residue on examination is always found to consist of that gas which was in excess of the due proportion. Thus, if 100 measures of *hydrogen* be mixed with 80 measures of *oxygen* and exploded, there will be 30 measures of gas left, which on examination will be found to be pure *oxygen;* or, on the other hand, if 150 measures of *hydrogen* be mixed with 50 measures of oxygen and exploded, there will be a residue of 50 measures, but in this case it will be found to consist of pure *hydrogen*. In all cases, if the gases are mixed otherwise than in the proportion of 1 of O to 2 of H, there will be a residue left, which will be found to be the gas which was in excess, and by exactly the amount of which it was in excess of the due proportion.

If, however, exactly 100 measures of *hydrogen* be mixed with 50 measures of oxygen, and exploded, no residue will be left; and if means have been taken to conduct the whole experiment at a temperature of above 100° C., it will be found that the resulting volume of gaseous water and steam from the explosion of the 150 volumes of the mixed gases will occupy exactly 100 volumes; so that two volumes of *hydrogen* combine with one volume of *oxygen* to form two volumes of aqueous vapour—

H | H	+	O	=	OH_2 | OH_2

The analytical composition of water, both by volume and weight, has been shown in Chapter VII., page 67; in fact nearly all the methods for the production of hydrogen are modes of analyzing water.

95. Absolutely pure water (such as we require for chemical purposes) is never to be found in nature; it is only obtainable by distillation, a method which was briefly described in Chapter I., page 10. For this purpose, on a large scale, a copper still is employed, with a copper or block tin worm for the condensation of the water. No lead must be employed, since lead is slightly soluble in pure water.

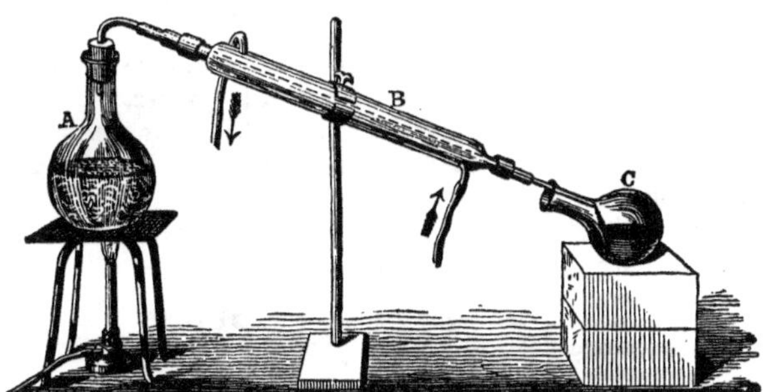

Fig. 40.

96. Properties.—Water, when pure, is a clear, transparent, odourless, and tasteless liquid. It has neither acid nor alkaline reaction, and may be taken as a type of a neutral substance; but its solvent powers (and this, with its neutrality, constitute its great value to the chemist) are greater than those of any other known liquid, and cause it never to be met with pure, unless prepared artificially.

It boils at 100° C., forming steam, and freezes at 0° C., but its behaviour with regard to heat is altogether anomalous. If ice at 0° C. be exposed to heat, contrary to the usual law, it contracts until it arrives at 4° C., when by the further application of heat, it follows the law and expands; so, on the contrary, warm water regularly contracts as it loses heat, until it reaches a temperature of 4° C., when it expands until it solidifies, and at the moment of solidification undergoes an expansion amounting to one-twelfth of its former bulk, so that ice is considerably lighter than water, and therefore floats upon it. Space will not allow me to dilate upon the importance, the uses, and the beauty of this arrangement; but the ice, staying at the surface, and being a bad conductor of heat, protects the waters beneath from further change, and the animal and vegetable life in the water from destruction during the long continued cold of our winter. Thus 4° C. (39·2° F.) may be taken as the point of greatest density of water.

At 4° C. one litre of water weighs 1000 grammes or one kilogramme; or, in English weights and measures, a cubic inch of water at 62° F. weighs 252·456 grains, and a cubic foot very nearly 1000 (more exactly 997) ounces avoirdupois.

97. Reactions.—1. By its action many metallic oxides are converted into hydrates*—

$$\mathbf{O}K_2 + \mathbf{O}H_2 = 2\mathbf{O}KH.$$
Potassic oxide. Water. Potassic hydrate.

$$\mathbf{Ba}O + \mathbf{O}H_2 = \mathbf{Ba}Ho_2.$$
Baric oxide. Water. Baric hydrate.

2. By its action on anhydrides it transforms them into acids—

* When a body is combined with water, it is usually termed a *hydrate*; when it is entirely free from water, it is said to be *anhydrous* (*ἀ, not; ὕδωρ, water*). This term, or rather that of "anhydride," is however usually restricted to those bodies which on the addition of water become acids.

$$\underset{\text{Nitric anhydride.}}{\mathbf{N}_2O_5} + \underset{\text{Water.}}{\mathbf{O}H_2} = \underset{\text{Nitric acid.}}{2\mathbf{N}O_2Ho.}$$

$$\underset{\text{Sulphuric anhydride.}}{\mathbf{S}O_3} + \mathbf{O}H_2 = \underset{\text{Sulphuric acid.}}{\mathbf{S}O_2Ho_2.}$$

$$\underset{\text{Phosphoric anhydride.}}{\mathbf{P}_2O_5} + 3\mathbf{O}H_2 = \underset{\text{Phosphoric acid.}}{2\mathbf{P}OHo_3.}$$

3. Many salts, when they crystallize, combine with a certain definite quantity of water, which is essential to the form they assume, and is called *water of crystallization.* Thus—

$\mathbf{Ba}Cl_2,2\mathbf{O}H_2$,................................*Baric chloride.*
$\mathbf{S}O_2Nao_2,10\mathbf{O}H_2$,..............................*Sodic sulphate.*

This last salt, like many others, when exposed to the atmosphere, parts with its *water of crystallization*, and falls into the condition of a white powder; it is then said to *effloresce.* Some salts, on the contrary, such as *potassic carbonate*, K_2CO_3, &c., when exposed to the atmosphere, absorb moisture, becoming damp and even liquid; such salts are said to *deliquesce.* That the *water of crystallization* is essential to the form of the salt, is shown not only by its losing its form when the water is driven off, but also by the fact that many salts crystallize in different forms, according to the amount of water they have absorbed.

On account of the extremely extensive solvent powers of water, it is never found pure in nature; that which falls in country districts, after long continued rain, is the purest, but even then it contains air and other gases in solution. That which falls in the neighbourhood of towns, or after long continued drought, always contains a number of impurities. Spring and river waters vary in impurity according to the district through which they flow. Thus, water from granite districts is the purest; that from the neighbourhood of towns, and from chalk and limestone districts, less pure. Some waters from the granite district of Scotland, and from the green sand of Surrey, contain as little as 4 or 5 per cent. of solid matter to the gallon; while that of the Thames as much as 21 grains per gallon. The purest natural water is perhaps that of Lake Loka, in the north of Sweden, which flows over hard impenetrable granite, and contains only $\frac{1}{20}$th of a grain per gallon. Next to that come the waters of Loch Katrine, with $2\frac{1}{2}$ grains of solid matter per gallon. The waters which spring from, or flow through, limestone districts are characterized by their bright, clear, sparkling appearance, due to the presence of lime and carbonic acid, and although, in a chemical sense, very impure, and unfitted for ordinary domestic purposes, are nevertheless exceedingly healthy and suitable for drinking. Water taken from surface wells in the neighbourhood of dwellings, farmyards, and graveyards, is frequently very impure, and even unwholesome to drink, from the amount of organic matter it contains.

98. Tests for Impurities.—The most common impurities of river or spring water are common salt (*sodic chloride*), NaCl, chalk (*calcic carbonate*), $CaCO_3$, gypsum (calcic sulphate), $CaSO_4$, with magnesic carbonate, and occasionally magnesic sulphate.

99. Test for Salt or any Soluble Chloride.—A solution of silver nitrate, $\mathbf{N}O_2Ago$, produces with salt or any soluble chloride, a *white curdy precipitate* of silver chloride (AgCl), which is insoluble in *nitric acid*, but is soluble in ammonia, NH_3—

$$NaCl + AgNO_3 = \mathbf{N}O_2Nao + AgCl.$$

EXP. 63.—Take two large test glasses, *perfectly* clean, and fill both with distilled water; into one drop a grain of common salt about the size of a pin's head, you cannot by the taste or appearance detect in which the salt is; but by the addition of two or three drops of silver nitrate to each it is at once made manifest. The one in which the salt was placed will become cloudy, while the other remains clear and bright. Take the cloudy solution, divide it into two parts, to one add a few drops of *nitric acid* (*hydrogen nitrate*) and stir, no effect will be produced; to the other add a few drops of *ammonia* and stir, the cloudiness will be immediately dissolved and disappear, and the solution become clear and bright.

100. Test for Calcic Carbonate.—Well boil the water, bubbles of gas (carbonic acid) will be seen to escape from it, while the calcic carbonate which it held in solution will be thrown down as a whitish-gray powder.

The *rationale* of this test is as follows:—Carbonate o *lime* is insoluble in water, but becomes soluble when the water holds *carbonic acid* in solution, as most spring waters do. By boiling this *carbonic* acid is expelled, and the carbonate of lime precipitated.

101. Test for any Lime Salt.—A solution of *ammonium oxalate* forms with *any* lime salt (and therefore with *calcic sulphate*) a white precipitate of *calcic oxalate,* which is insoluble in *acetic acid.*

102. Test for Sulphuric Acid.—A solution of *baric chloride* or *baric nitrate* forms with *sulphuric acid* a white precipitate of baric sulphate, which is insoluble in all acids—

SO_4Cao	+	$BaCl_2$	=	$CaCl_2$	+	SO_4Bao
Calcic sulphate.		Baric chloride.		Calcic chloride.		Baric sulphate.

Hydroxyl.

103. Symbol, $(HO)_2$, or H_2O_2, or Ho_2, or $\left\{\begin{matrix}\mathbf{O}H \\ \mathbf{O}H\end{matrix}\right.$ Graphic formula, (H)-(O)-¦-(O)-(H). Molecular weight = 34.

104. Hydroxyl, hydric dioxide, or peroxide of hydrogen, the second compound of oxygen and hydrogen, was discovered by Thénard in 1818.

The molecule of hydroxyl differs from that of water, in having an additional atom of oxygen united to the two atoms of hydrogen. So that its molecule is made up of two atoms of the compound radical, $\mathbf{O}''H$, (H)—(O)- or Ho'.

105. Preparation.—The preparation of hydroxyl is an indirect process, and is attended with so much difficulty that but few chemists have isolated it.

1. By passing a current of *carbonic anhydride* through water in which *baric peroxide* is suspended—

$$\left\{\begin{matrix}\mathbf{O}\\ \mathbf{O}\end{matrix}\right.Ba'' \quad \mathbf{C}O_2 \quad + \quad \mathbf{O}H_2 \quad = \quad \mathbf{C}OBao'' \quad + \quad \left\{\begin{matrix}\mathbf{O}H\\ \mathbf{O}H\end{matrix}\right.$$

Baric peroxide. Carbonic anhydride. Water. Baric carbonate. Hydroxyl.

$$[BaO_2 + CO_2 + H_2O = BaCO_3 + 2HO.]$$

106. Properties.—Hydroxyl so obtained is a colourless liquid of a syrupy consistence, an astringent taste, and an odour resembling that of chlorine very much diluted. Its specific gravity is 1·453. At a temperature of – 30° C. (– 22° F.) it remains liquid, but at a higher temperature it decomposes. The compound is a very unstable one. At 20° C. (68° F.) it evolves oxygen somewhat rapidly, but as it approaches 100° C., it is decomposed with rapidity, and even with explosive violence. It is soluble in water, and is then much more stable, especially if a little sulphuric acid is added to it.

107. Hydroxyl, like chlorine, has powerful bleaching properties; it destroys litmus, and nearly all vegetable colours; if a drop be placed on the hand, it produces at once a white spot. Could it be prepared easily and economically, it would be invaluable as a bleaching and oxidating agent.

108. Reactions.—1. By heat it is decomposed into water and oxygen—

$$4\left\{\begin{matrix}\mathbf{O}H\\ \mathbf{O}H\end{matrix}\right. \quad = \quad 2\mathbf{O}H_2 \quad + \quad O_2.$$

Hydroxyl. Water. Oxygen.

This forms an easy method of analyzing the hydroxyl. A given weight of it is placed in a retort with ten or twelve times its bulk of water, and distilled over. When it rises to boiling point, oxygen is given off freely, the gas is collected over mercury and allowed to cool; from its volume its weight can be easily calculated, and it is found that for every 16 parts of O given off, 18 of water remain behind, so if 18 of water consist of 2 of H + 16 of O, *hydroxyl* must consist of 2 of H + 2 × 16 or 32 of O, and therefore its constitution must be H_2O_2.

2. If hydroxyl be mixed with potassic iodide the iodine is set at liberty, potassic hydrate being formed—

$$\underset{\text{Potassic iodide.}}{2KI} + \underset{\text{Hydroxyl.}}{\left\{\begin{matrix}\mathbf{O}H\\ \mathbf{O}H\end{matrix}\right.} = \underset{\text{Potassic hydrate.}}{2\mathbf{O}KH} + \underset{\text{Iodine.}}{I_2}.$$

If a cold solution of starch be taken into which a small quantity of potassic iodide is put, and a few drops of hydroxyl be added, the above reaction will be rendered evident by the blue colour, indicative of the existence of iodine in the presence of starch.

3. It is a powerful oxidizing agent; it converts the black plumbic sulphide into white plumbic sulphate, a property which has been turned to account in cleaning oil paintings, in which the white lead has become blackened by exposure to air containing hydrogen sulphide—

$$\underset{\text{Plumbic sulphide.}}{\mathbf{Pb}S''} + \underset{\text{Hydroxyl.}}{4\left\{\begin{matrix}\mathbf{O}H\\ \mathbf{O}H\end{matrix}\right.} = \underset{\text{Plumbic sulphate.}}{\mathbf{S}O_2Pbo''} + \underset{\text{Water.}}{4\mathbf{O}H_2}.$$

Many of the metals, such as gold, platinum, and silver, and the peroxides of manganese and lead, especially in a fine state of subdivision, decompose hydroxyl into oxygen and water, by a catalytic action, but themselves undergo no change; while some of the protoxides, as the protoxide of lead, effect this decomposition, but are themselves raised to a higher degree of oxidation. The oxides of gold, silver, and platinum, not only decompose hydroxyl, but are themselves also reduced to the metallic state.

Compounds of Oxygen and Hydroxyl with Chlorine.

109. **Chlorine** forms with *oxygen* and *hydroxyl* many compounds, but none of them can be obtained by direct combination. The following are all that are known:—

110. **Hypochlorous Anhydride.**—Symbol, $\mathbf{O}Cl_2$ [Cl_2O]. Graphic formula, (CL)-(O)-(CL). Atomic weight = 87. Molecular volume, ☐☐. Relative weight, 43·5. Sp. gr. (theoretic), 3·005. Boiling point about 20° C. (68° F.)

111. Preparation.—By passing chlorine over perfectly dry mercuric oxide at a low temperature—

$$\underset{\text{Mercuric oxide.}}{2\mathbf{Hg}O} + \underset{\text{Chlorine.}}{2Cl_2} = \underset{\text{Mercuric oxychloride.}}{\left\{\begin{array}{l}\mathbf{Hg}Cl\\ O\\ \mathbf{Hg}Cl\end{array}\right.} + \underset{\text{Hypochlorous anhydride.}}{\mathbf{O}Cl_2}.$$

The chlorine should be passed from the generating flask through a wash bottle containing water to free it from hydrochloric acid, then through a U-tube, filled with pumice stone, moistened with sulphuric acid, to dry it; it is then passed through a tube filled with mercuric oxide, and finally received into a bent receiver, which is surrounded by a freezing mixture of pounded ice and salt. In this receiver the hypochlorous anhydride is condensed into a deep red liquid which emits a vapour of a deeper colour than chlorine, and of a strong, suffocating, chlorous odour. The warmth of the hand even decomposes this vapour with explosive violence, resolving it into chlorine and oxygen; two volumes of the vapour producing two volumes of chlorine and one of oxygen.

112. Chlorous Anhydride.—Symbol, $\left\{\begin{array}{l}\mathbf{O}Cl\\ O\\ \mathbf{O}Cl,\end{array}\right.$ $[Cl_2O_3]$.

Graphic formula, (Cl)–(O)–(O)–(O)–(Cl). Molecular weight $= 119$. Molecular volume (anomalous), $\boxed{\;\cdot\;\cdot\;\cdot\;}$. Relative weight, 37·7. Sp. gr., 2·46.

113. Preparation.—By gently heating in a water bath a mixture of potassic chlorate, nitric acid, and arsenious acid, the reaction takes place in four stages, as follows:—

$$1.\quad \underset{\text{Potassic chlorate.}}{\left\{\begin{array}{l}\mathbf{O}Cl\\ \mathbf{O}Ko\end{array}\right.} + \underset{\text{Nitric acid.}}{\mathbf{N}O_2Ho} = \underset{\text{Chloric acid.}}{\left\{\begin{array}{l}\mathbf{O}Cl\\ \mathbf{O}Ho\end{array}\right.} + \underset{\text{Potassic nitrate.}}{\mathbf{N}O_2Ko.}$$

$$2.\quad \underset{\text{Arsenious acid.}}{\mathbf{As}Ho_3} + \underset{\text{Nitric acid.}}{\mathbf{N}O_2Ho} = \underset{\text{Nitrous acid.}}{\mathbf{N}OHo} + \underset{\text{Arsenic acid.}}{\mathbf{As}OHo_3}$$

3. $$\left\{\begin{matrix}\mathbf{O}\mathrm{Cl}\\ \mathbf{O}\mathrm{Ho}\end{matrix}\right. + \mathbf{N}\mathrm{OHo} = \mathbf{O}\mathrm{ClHo} + \mathbf{N}\mathrm{O_2Ho}.$$

Chloric acid. Nitrous acid. Chlorous acid. Nitric acid.

4. $$2\mathbf{O}\mathrm{ClHo} + \mathbf{O} = \left\{\begin{matrix}\mathbf{O}\mathrm{Cl}\\ \mathrm{O}\\ \mathbf{O}\mathrm{Cl}\end{matrix}\right. + \mathbf{O}\mathrm{H_2}.$$

Chlorous acid. Chlorous anhydride. Water.

114. Chlorous Anhydride is a very dangerous compound to prepare, as at a temperature exceeding 55° C. (131° F.) it decomposes with a violent explosion. Contact with most combustible non-metallic elements, as S, P, Se, &c., also causes its violent decomposition. Most of the metals have no action on it, but mercury absorbs it completely. The gas is soluble in about $\frac{1}{6}$th of its bulk of water, when it forms chlorous acid.

115. Chloric Peroxide.—Symbol, $\left\{\begin{matrix}\mathbf{O}\mathrm{Cl}\\ \mathrm{O}\\ \mathrm{O}\\ \mathbf{O}\mathrm{Cl}\end{matrix}\right.$ $[\mathrm{ClO_2}]$. Graphic formula, (Cl)-(O)-(O)-(O)-(O)-(Cl). Molecular weight, 135. Relative weight, 33·75. Boiling point, 20° C. (68° F.)

116. Preparation.—It is obtained by acting on fused potassic chlorate (broken into coarse fragments), with about two-thirds of its weight of sulphuric acid. The action requires to be assisted sometimes by a gentle heat; the reaction is represented by the following equation:—

$$3\left\{\begin{matrix}\mathbf{O}\mathrm{Cl}\\ \mathrm{OKo}\end{matrix}\right. + 2\mathbf{O}\mathrm{O_2Ho_2}\mathbf{O} = \left\{\begin{matrix}\mathbf{O}\mathrm{Cl}\\ \mathrm{O}\\ \mathbf{O}\mathrm{Ko}\end{matrix}\right. + 2\mathbf{O}\mathrm{O_2HoKo} + \mathbf{O}\mathrm{H} + \left\{\begin{matrix}\mathbf{O}\mathrm{Cl}\\ \mathrm{O}\\ \mathrm{O}\\ \mathbf{O}\mathrm{Cl}\end{matrix}\right.$$

Potassic calorate. Sulphuric acid. Potassic perchlorate. Hydric potassic sulphate. Water. Chloric peroxide.

$$[3\mathrm{KClO_3} + 2\mathrm{H_2SO_4} = \mathrm{KClO_4} + 2\mathrm{KHSO_4} + \mathrm{H_2O} + 2\mathrm{ClO_2}.]$$

117. Properties.—This compound is gaseous at ordinary temperatures, but by a slight increase of pressure, or at a temperature of −20° C. (−4° F.), it is reduced to a red liquid, which is, however, exceedingly unstable and liable to explosion. The gas is of a deeper colour than chlorous anhydride, and of a similar but less irritating odour. In the dark it remains unaltered, but sunlight gradually decomposes it into its constituent gases. The gas explodes at a temperature of from 60° to 63° C. (140° to 145° F.), and therefore requires great care in its preparation. Water dissolves about 20 times its bulk of the gas, and forms a yellow solution which possesses powerful bleaching properties.

118. As **Chloric peroxide** acts powerfully on mercury, it has to be collected by downward displacement. It is not possessed of acid properties, but is rapidly absorbed by alkaline solutions. Mere contact with many combustibles determine its explosion.

EXP. 64.—Take equal quantities (from ½ oz. to 1 oz. of each is quite sufficient) of potassic chlorate and loaf sugar; powder them separately, and mix with a spatula on paper; place the mixture in an earthenware dish (a flower-pot saucer will do very well); add but a drop of sulphuric acid from the end of a glass rod dipped in the acid, the chloric peroxide will be liberated, and will be decomposed instantaneously on contact with the combustible matter, and sufficient heat will be generated to cause the whole mass to burst into flame, and deflagrate brilliantly.

Other oxides of chlorine have been obtained, but their constitution and their properties have not yet been accurately determined.

119. Hypochlorous Acid.—Symbol, **O**ClH, or ClHo [HClO]. Molecular and atomic weight, 52·5. Graphic formula Ⓗ—Ⓞ—Ⓗ

120. Preparation.—1. By the action of chlorine upon mercuric oxide, in the presence of water—

$$2\mathbf{Hg}O + \mathbf{O}H_2 + 2Cl_2 = \left\{\begin{matrix}\mathbf{Hg}Cl \\ O \\ \mathbf{Hg}Cl\end{matrix}\right. + 2ClHo.$$

Mercuric oxide. Water. Chlorine. Mercuric oxychloride. Hypochlorous acid.

2. By the action of water on hypochlorous anhydride—

$$\mathbf{O}Cl_2 + \mathbf{O}H_2 = 2ClHo.$$

Hypochlorous anhydride. Water. Hypochlorous acid.

Water dissolves about 200 times its bulk of gaseous hypochlorous anhydride, and forms with it a pale yellow solution, which has an acrid but not a sour taste. Like all the compounds of chlorine, it is exceedingly unstable; when exposed to the light it is rapidly decomposed, chlorine being given off in bubbles. It is a powerful oxidizing agent. A solution of hypochlorous acid is decomposed rapidly by any of the following elements:—C, I, S, Se, P, As, and Sb (the latter should be in the state of fine powder). And the corresponding acids are formed—carbonic, iodic, sulphuric, selenic, phosphoric, arsenic, and antimonic.

When brought into contact with silver, the chlorine is absorbed, and oxygen set free, while copper and mercury absorb both oxygen and chlorine, and form oxychlorides; but its most important application is its bleaching power, which is double that of chlorine.

When it combines with the alkalies or earths, it forms the salts called *hypochlorites*. These salts are decomposed by the weakest acids, even the carbonic; and the chlorine set free shows its presence by its usual bleaching effect on vegetable colours.

The most important of these compounds of chlorine, with regard to bleaching power, are those with the alkali metals, and notably with that of lime, called *chloride of lime*, but most probably a *hypochlorite of lime*, its formation seeming to be represented by the following equation.

$$\mathbf{Ca}Ho_2 + Cl_2 = \mathbf{Ca}(OCl)Cl + \mathbf{O}H_2.$$

Calcic hydrate. Chlorine. Bleaching powder. Water.

By the action of acids, this compound yields free chlorine—

$$\mathbf{Ca}(OCl)Cl + \mathbf{S}O_2Ho_2 = \mathbf{S}O_2Cao'' + \mathbf{O}H_2 + Cl_2.$$

Bleaching powder. Sulphuric acid. Calcic sulphate. Water. Chlorine.

121. Chlorous Acid.—Symbol, $\mathbf{O}ClHo$ or $\left\{\begin{matrix}\mathbf{O}Cl\\ \mathbf{O}H\end{matrix}\right.$ [$HClO_2$]. Molecular weight, 68·5.

122. Preparation.—It is formed by the action of water on chlorous anhydride :—

$$\left\{\begin{matrix}\mathbf{O}Cl\\ O\\ \mathbf{O}Cl\end{matrix}\right. + \mathbf{O}H_2 = 2\left\{\begin{matrix}\mathbf{O}Cl\\ \mathbf{O}H\end{matrix}\right.$$

Chlorous anhydride. Water. Chlorous acid.

Like all the compounds of chlorine, it possesses great bleaching and oxidating powers; it acts slowly on bases, forming a series of salts called **chlorites.** Most of the chlorites are deliquescent. They may be expressed by the general formula $M'ClO_2$. They are very unstable, being decomposed by the weakest acids, even by carbonic acid.

123. Chloric Acid.—Symbol, $\left\{\begin{matrix}\mathbf{O}Cl\\ \mathbf{O}Ho\end{matrix}\right.$ or $\left\{\begin{matrix}\mathbf{O}Cl\\ O\\ \mathbf{O}H\end{matrix}\right.$ or [$HClO_3$]. Molecular weight, 84·5.

124. Preparation.—1. By the action of diluted hydrofluosilicic acid on potassic chlorate :—

$$2\left\{\begin{matrix}\mathbf{O}Cl\\ \mathbf{O}Ko\end{matrix}\right. + SiH_2F_6 = 2\left\{\begin{matrix}\mathbf{O}Cl\\ \mathbf{O}Ho\end{matrix}\right. + SiK_2F_6.$$

Potassic chlorate. Hydrofluosilicic acid. Chloric acid. Potassic silicofluoride.

$$[2KClO_3 + H_2SiF_6 = 2HClO_3 + K_2SiF_6.]$$

2. By the action of dilute sulphuric acid upon baric chlorate :—

$$\left\{\begin{matrix}\mathbf{O}Cl\\ O\\ Bao''\\ O\\ \mathbf{O}Cl\end{matrix}\right. + \mathbf{O}O_2Ho_2 = 2\left\{\begin{matrix}\mathbf{O}Cl\\ \mathbf{O}Ho\end{matrix}\right. + SO_2Bao''.$$

Baric chlorate. Sulphuric acid. Chloric acid. Baric sulphate.

$$[BaCl_2O_6 + H_2SO_4 = 2HClO_3 + BaSo_3.]$$

The reaction in both cases is similar in throwing down the base of the chlorate in an insoluble form, which may then be separated by filtration,* leaving a weak solution of the acid.

This acid solution may then be evaporated over the water bath, at a temperature not exceeding 38° C. (100·4° F.), till it attains the consistence of syrup.

125. Properties.—It has a powerfully acid taste, and a faint chlorous smell. It is immediately decomposed by organic matter, and when concentrated will set fire to paper.

At a temperature a little above 38° C. (100° F.) it is decomposed into perchloric acid, water, chlorine, and oxygen—

$$3\left\{\begin{matrix}\mathbf{O}Cl\\ \mathbf{O}Ho\end{matrix}\right. = \left\{\begin{matrix}\mathbf{O}Cl\\ O\\ \mathbf{O}Ho\end{matrix}\right. + \mathbf{O}H_2 + Cl_2 + 2O_2.$$

Chloric acid. Perchloric acid. Water. Chlorine. Oxygen.

$$[3HClO_3 = HClO_4 + H_2O + Cl_2 + 2O_2.]$$

It also undergoes spontaneous decomposition gradually in diffused daylight.

Note.—Both chloric and perchloric acids, if allowed to touch the skin, produce wounds which are very painful and difficult to heal.

126. Preparation and Properties of the Chlorates.—The chief use of *chloric acid* (hydric chlorate) is in the preparation of the metallic chlorates. This may be done by neutralizing the acid by means of the carbonate or oxide of the metal, thus:—

$$[BaO + 2HClO_3 = Ba2ClO_3 + H_2O.]$$
$$[PbO + 2HClO_3 = Pb2ClO_3 + H_2O.]$$

The most important of the chlorates is that of potassium, which is prepared as follows:—

A current of chlorine is passed into a saturated solution of potassic hydrate—

* The filtration must be conducted through a plug of gun cotton or asbestos, as the chloric acid would be decomposed by contact with the filter paper

$$6OKH + 3Cl_2 = 5KCl + \left\{\begin{matrix} OCl \\ OKo \end{matrix}\right. + 3OH_2.$$

Potassic hydrate.	Chlorine.	Potassic chloride.	Potassic chlorate.

This process is, however, not economical in practice, five-sixths of the potassium being converted into its chloride, a comparatively cheap and useless salt. It is therefore customary in manufacturing potassic chlorate on a large scale to use calcic chlorate instead of potassic hydrate.

By the addition of potassic chloride to the calcic chlorate, potassic chlorate is formed, which can be separated from the calcic chloride by crystallization—

$$\left\{\begin{matrix} OCl \\ O \\ Cao'' \\ O \\ OCl \end{matrix}\right. + 2KCl = 2\left\{\begin{matrix} OCl \\ OKo \end{matrix}\right. + CaCl_2.$$

Note.—Calcic chlorate can be easily made by passing a current of chlorine through a boiling solution of milk of lime.

The chlorates are all soluble in water; they are all decomposed by heat, giving up either a part or all of their oxygen, and leaving the metal in the form of an oxide or chloride; they all deflagrate when thrown on lighted charcoal; they all give up their oxygen readily to all combustible bodies, especially when heated, combining with some, as phosphorous sulphur and antimony, with sufficient violence to cause an explosion. On this account they are much used for fireworks and coloured fires, the metal with which they are combined imparting different colours to the flame, thus—

Strontic	chlorate or	nitrate	imparts to flame	a bright	red	tint.
Baric	,,	,,	,,	,,	green	,,
Potassic	,,	,,	,,	,,	violet	,,
Sodic	,,	,,	,,	,,	yellow	,,
Cupric	,,	,,	,,	,,	blue	,,

This eagerness of the chlorates to enter into active combustion was illustrated in Exp. 64. It frequently happens that mere friction is sufficient to produce a violent detonation, for example—

EXP. 65.—If the same bodies be powdered separately, then carefully mixed together on a card, with as little friction as possible, wrapped up tightly in paper, and struck a smart blow with a hammer, a loud explosion will be produced.

EXP. 66.—A mixture which detonates powerfully when struck or rubbed, may be obtained by powdering separately equal quantities of antimonious sulphide (Sb_2S_3) and potassic chlorate ($KClO_3$), and mixing them lightly and cautiously with a feather. *Note.*—Do not use more than one grain of each.

Note.—The union of chlorine with hydrogen is exceeding stable, but all the compounds of chlorine with oxygen and hydroxyl are easily decomposed.

127. Perchloric Acid.—Symbol, $\left\{\begin{array}{l}\mathbf{O}Cl\\ O\\ \mathbf{O}Ho\end{array}\right.$ $[HClO_4]$. Molecular weight, 100·5. Specific gravity, 1·782.

CHAPTER XII.

Boron—History—Its Occurrence in Nature—Its Allotropic Modifications—Boric Anhydride—Boric Acids.

128. Boron.—Symbol B. Atomic weight = 11. Probable molecular weight = 22. Sp. gr. of crystals, 2·68. Atomicity,‴. Evidence of atomicity:—

Boric chloride,.. $\mathbf{B}'''Cl$
Boric fluoride,.. $\mathbf{B}'''F_3$.

129. History.—Boron was discovered by Davy in 1807, and about the same time, independently, by Gay Lussac and Thenard.

130. Occurrence.—It is never found pure in nature, being always in combination with oxygen as *boric anhydride*, either free, or more generally in combination with some of the metals. Its most important salt is *borax*, a compound of the base *soda* with *boric anhydride.*

131. Allotropic Modifications.—*Boron* differs from *carbon* and *silicon* in its atomicity, being a *triad*, while those elements are each of them *tetrads;* but it resembles them in being capable of existing in three allotropic forms:—(1.) The *amorphous;* (2.) The semi-crystalline or graphitic; and (3.) The crystalline variety. It also presents great analogy with silicon in its properties and mode of combination.

132. Preparation.—(1.) *Amorphous Boron*—Boric anhydride is ignited with sodium in a covered crucible. The oxygen leaves the boron and unites with the sodium, forming sodic oxide, which may be removed by treatment with hot water and dilute hydrochloric acid. (See Miller's *Elements of Chemistry,* vol. II., p. 252.)

$\mathbf{B}_2O_3$	+	$3Na_2$	=	$3\mathbf{O}Na_2$	+	B_2.
Boric anhydride.		Sodium.		Sodic oxide.		Boron.

Or by passing boric chloride over heated potassium—

$2\mathbf{B}Cl_3$	+	$3K_2$	=	$6KCl$	+	B_2.
Boric chloride.				Potassic chloride.		

133. Properties.—Amorphous boron, as thus obtained, is a dull olive green powder, which, however, becomes denser and darker when heated in vessels from which the air is excluded. It is very slightly soluble in pure water. It is not very readily oxidizable, except when heated, and it then burns either in air or oxygen gas, with a reddish light and vivid scintillations. It also deflagrates powerfully if mixed with potassic nitrate and heated to redness.

It decomposes hot sulphuric acid, nitric acid, and also carbonates, sulphates, and nitrates of the alkaline metals (K, Na):—

B_2	+	$3\mathbf{O}O_2Ho_2$	=	$\mathbf{B}_2O_3$	+	$3\mathbf{O}H_2$	+	$3\mathbf{O}O_2$.
		Sulphuric acid.		Boric anhydride.		Water.		Sulphurous anhydride.

B_2	+	$6\mathbf{N}O_2Ho$	=	$2BHo_3$	+	$3N^{iv}O_4$.
		Nitric acid.		Boric acid.		Nitric peroxide.

B_2	+	$3\mathbf{C}ONao_2$	=	$2\mathbf{B}Nao_3$	+	$3C''O$.
		Sodic carbonate.		Trisodic borate.		Carbonic oxide.

Boron is the least abundant of the non-metals; it is met

with in but few places, but on account of the powerful properties as a flux which borax possesses, it is of great chemical importance.

134. Boric Anhydride.—Also called *boracic anhydride.* Symbol, $\mathbf{B}_2O_3$. Atomic and molecular weight = 70. Sp. gr., 1·83. Graphic formula, $O{=}B{-}O{-}B{=}O$

135. Preparation.—By fusing boric acid at a red heat—

$$\underset{\text{Boric acid.}}{2\mathbf{B}Ho_3} = \underset{\text{Boric anhydride.}}{\mathbf{B}_2O_3} + \underset{\text{Water.}}{3\mathbf{O}H_2}.$$

The acid merely fuses, and forms on cooling a glassy, hard, ringing mass. A certain amount of loss from volatilization always takes place.

Boric Acid.

136. Boracic Acid, Orthoboric Acid.—Symbol, BHo_3. Atomic and molecular weights, 62. Sp. gr., 1·479.

Graphic formula, B bonded to three O–H groups (H–O–B(–O–H)–O–H)

137. Occurrence.—It is found in the mineral called *tincal,* an acid *borate of sodium* ($\mathbf{B}_4O_5Nao_2,10\mathbf{O}H_2$), obtained from Thibet, and in *boracite,* a *borate of magnesia,* found in Saxony; but its most abundant source is the Maremma of Tuscany, where it is found, in an uncombined state, associated with hydrogen sulphide; it is mixed with the steam which issues from the *fumeroles* or *soffioni.*

The steam and vapours from these *fumeroles* are conducted into artificial basins or lagoons, the waters of which on evaporation yield a crude boracic acid, from which the greater part of the borax of commerce is manufactured.

For a description of the process see Miller's *Chemistry*, vol. II., page 253.

The commercial acid is purified by adding to it sodic carbonate as long as effervescence occurs, when *borax* or *sodic borate* is formed, which may be obtained by crystallization.

The pure *boric acid* may then be obtained by treating a hot saturated solution of *borax* with either *hydrochloric* or *sulphuric* acids, when the *boric acid* is set free and *sodic chloride* or *sodic sulphate* is formed. These being very soluble, while *boric acid* is sparingly so, the latter crystallizes out on cooling in the form of pearly looking scales, with a greasy feeling to the touch—

$\mathbf{B}_4O_5Nao_2$*	+	2HCl	+	$5\mathbf{O}H_2$	=	$4\mathbf{B}Ho_3$	+	2NaCl.
Borax.		Hydrochloric acid.		Water.		Boric acid.		Sodic chloride.

138. Properties.—The crystals of *boric acid* effloresce and lose water at a slight heat; but at a temperature of 100°C. it becomes, by the loss of water, converted into *metaboric acid*, thus—

$\mathbf{O}Ho_3$	=	$\mathbf{O}OHo$	+	$\mathbf{O}H_2$.
Boric acid.		Metaboric acid.		Water.

By an increase of temperature all the water is driven off, and it becomes converted into boric anhydride. *Boric acid* imparts to its compounds the property of fusing very readily. This is in fact its chief use. Many of the borates, especially *sodic borate*, are largely used as *fluxes* in the reducing of metals from their ores, and in many processes of the arts, as in the glazing of porcelain.

Most borates fuse to transparent glasses. Alkaline borates, when fused with certain metallic oxides, take

* The graphic formula for borax is

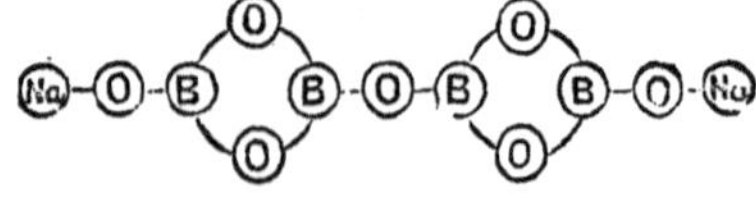

them up and dissolve them, forming a double borate of the alkali and the metallic oxide which are marked by characteristic colours. Hence the great use of *borax* in analysis and in blowpipe reactions.

CHAPTER XIII.

Carbon—Preparation—Allotropic Forms—Combinations—Preparation and Properties of Carbonic Oxide—Carbonic Anhydride, its Preparation and Properties.

139. Carbon.—Symbol, C. Atomic weight = 12. Atomicy $0''$ and 0^{iv}. Specific gravity varies; in its most dense form, that of the *diamond*, the sp. gr. = 3·5.

140. Occurrence.—Most abundant in all nature, being found to a slight extent in the air and most natural water, somewhat abundantly in many of the rocks, and very largely in *all* animals and vegetables. It is, however, generally found in combination.

Carbon, like boron, occurs in three distinct forms, or allotropic modifications—viz., the *crystalline*, the *graphitic*, and the *amorphous*.

141.—1. Crystalline Carbon, as the *diamond.* These have never been produced artificially. They are found native in various parts of the earth, and are noted for their intense hardness, the high lustre they are capable of receiving, and their high refractive power on light; these qualities, combined with their rarity, cause them to be highly estimated as jewels. If a diamond be heated to redness, and plunged in oxygen gas, it swells up, and becomes converted into a mass resembling coke, which turns away, and leaves as the product of combustion pure *carbonic anhydride*, CO_2. The diamond crystallizes in forms belonging to the cubical system (1st system). It is the very purest form of carbon.

142.—2. Graphitic Carbon, as "*graphite*" (from γράφω,

I write). It is found among the older metamorphic rocks in many parts of the earth, and is probably altered vegetation. In England it has been generally obtained from the mine of Borrowdale, in Cumberland, but that is now nearly exhausted.

143. Properties.—Graphite, plumbago, or black lead, may be known by its peculiar metallic gray lead-like lustre, its power of conducting heat and electricity, its specific gravity (2·15 to 2·35), its softness, the marks it leaves on paper, and its crystallizing in hexagonal tables belonging to the rhombohedral system. It is never found perfectly pure, as when burned in oxygen it always leaves a small amount (from 2 to 5 per cent.) of ash.

144. Uses.—For the manufacture of lead pencils; when mixed with fire-clay it makes crucibles, largely used by the metallurgist; as a polish for iron surfaces to prevent them from rusting; and also to lubricate machinery.

145. Preparation.—It can be artificially prepared by allowing iron which has been heated with excess of carbon to cool slowly. Small quantities of *graphite*, in the shape of minute black scales, are always found disseminated through the mass of gray cast iron. A hard variety of *imperfect graphite* is also found coating the inside of the hottest parts of the retorts used in the manufacture of coal gas. It is this *graphite* which is used as the negative element in Bunsen's batteries, and also as the electrodes in the production of the electric light.

146.—3. Amorphous Carbon, as *charcoal, coke, soot, &c.* *Carbon* is never found naturally in the amorphous condition. *Charcoal* is of two kinds—*wood charcoal* and *animal charcoal*, or bone black.

147. Wood Charcoal is prepared either by distilling wood in covered crucibles, when all the volatile products, as *acetic acid*, &c., and a small portion of the carbon, pass over, leaving the greater part of the C behind as charcoal; or by heaping up billets of wood around a central heap of brushwood, and then covering the whole mass with powdered charcoal, earth, and clay, so as to exclude the air, with the exception of a small opening through which lighted faggots can be introduced to fire the brushwood in the centre, and through which the admission of air can be regulated. When tar ceases to be formed, it is covered up and left to cool. When cold, the charcoal is fit for use. The quantity and quality of the charcoal produced depends much on the nature of the wood used, and the rate of combustion.

148. Animal Charcoal, or bone black, is made by burning bones, or any animal substance, in a closed vessel.

Exp. 67.—Nearly fill a tube with dry ammonia or hydrochloric acid gas over mercury; then heat a piece of charcoal about the size of a bean to redness, to expel all the moisture, and plunge it, while still glowing, beneath the surface of the mercury; after keeping it there a few seconds, allow it to ascend into the gas in the tube. Note that the volume of the gas diminishes, and the mercury rises in the tube to supply its place. Any gas might be used, but the more soluble ones are absorbed most rapidly. The object in exposing the charcoal to a red heat is to expel all moisture or gas with which it may happen to be saturated.

If the charcoal be exposed to the gas for twenty-four hours, at the ordinary temperature and pressure of the atmosphere, its power of absorption for the different gases may be expressed by the following table:—

1 vol.	of charcoal	absorbs about	90	vols.	of *ammonia.*
1 ,,	,,	,,	85	,,	of hydrochloric acid.
1 ,,	,,	,,	65	,,	of sulphurous anhydride.
1 ,,	,,	,,	55	,,	of hydrogen sulphide.
1 ,,	,,	,,	35	,,	of carbonic anhydride.
1 ,,	,,	,,	9·25	,,	of oxygen.
1 ,,	,,	,,	1·25	,,	of hydrogen.

If a piece of charcoal saturated with *hydrogen sulphide* be brought in contact with free *oxygen,* an explosion takes place. It is to this power of absorption which charcoal possesses that it owes its deodorizing, disinfecting, and decolorizing properties. The oxygen gets absorbed in the pores of the charcoal, and is thus brought into close contact with the other gases which are absorbed, which being thus burnt or oxidized are deprived of their odour.

149. Properties.—The most remarkable property of amorphous carbon, in the form of charcoal (especially when powdered), is its power of absorbing gases and vapours, and especially organic colouring matter.

Animal charcoal possesses this property also of absorbing and deodorizing gases, but in a much less degree; but, on the other hand, it possesses in a remarkable manner the power of throwing down and decolorizing all organic substances, as indigo, logwood, &c., a property which renders it very valuable to the sugar refiner, and also for the purpose of filtering out organic impurities from water.

150.—Soot and **Lamp Black** are neither of them pure forms of carbon. The former, which is condensed smoke, is valuable as a manure on account of the *ammonia* it contains. The latter, which is obtained by burning bodies rich in C in a limited supply of air, and collecting and condensing the soot, is most probably a mixture of carbon and several hydrocarbons. Its chief use is in the manufacture of Indian ink and other black pigments, while the coarser varieties are used for printers' ink.

151.—Coke is the residue left when coal has been exposed to a great heat in a closed retort or oven, into which no air has been admitted, but from which the volatile products can escape freely. Coke is very nearly pure carbon. It does not ignite so easily as coal, but it gives out more heat than an equal weight of coal, and with little or no smoke.

152. Combinations.—*Carbon* combines with—

Oxygen to form *carbonic anhydride*, CO_2.

Oxygen to form *carbonic oxide*, CO.

Hydrogen, *chlorine*, and *nitrogen*, in a large number of proportions, forming an extensive series of compounds; but as all these belong to *Organic Chemistry*, they will be treated of in the volume on that subject.

Sulphur to form *carbonic disulphide*, CS_2''.

153. Carbonic Anhydride.—Symbol, CO_2. Molecular volume, $\square\square$. Molecular weight = 44. 1 litre weighs 22 criths. Relative weight therefore = 22. Specific gravity, 1·52. Fuses at – 57°. Boils below its melting point.

154. Synonyms.—Carbonic anhydride. Carbonic dioxide. Carbonic acid. Fixed air.

155. History.—Discovered by Dr. Black in 1757; and from the fact of his finding it as a fixed or solid constituent of limestone, and from its becoming fixed or absorbed by solutions of the caustic alkalies, he called it "fixed air."

156. Occurrence.—In a free state in the atmosphere, and in solution in water.

157. Formation.—By the burning of carbon and all carbonaceous substances in the air or in oxygen. In the respiration of man and animals; in all the various processes of decay and fermentation which are going on around us; in the burning of lime in the lime-kiln. From volcanoes and volcanic action, carbonic anhydride is largely evolved.

158. Preparation.—1. It can be artificially prepared by burning carbon in air or oxygen—

$$C + O_2 = \mathbf{C}O_2$$

Carbonic anhydride.

2. By the action of an acid on metallic carbonates—

$$\mathbf{C}OKo_2 + \mathbf{S}O_2Ho_2 = \mathbf{C}O_2 + \mathbf{O}H_2$$

Potassic carbonate. Sulphuric acid. Carbonic anhydride. Water.

$$[KCO_3 + H_2SO_4 = CO_2 + H_2O$$

$$+ \mathbf{S}O_2Ko_2$$

Potassic sulphate.

$$+ KSO_4.]$$

$$\mathbf{C}OCao'' + 2HCl = \mathbf{C}O_2 + \mathbf{O}H_2 + \mathbf{Ca}Cl_2$$

Calcic carbonate. Hydrochloric acid. Carbonic anhydride. Water. Calcic chloride.

The latter method is much the best one for producing carbonic anhydride in any quantity sufficient for the examination of its properties.

Marble, limestone, Iceland spar, chalk, shells—as oyster and mussel shells, &c.—are all of them carbonate of lime, and may be represented by the formula $\mathbf{C}OCao''$

[$CaCO_3$], and all yield carbonic acid when acted on by a sufficiently strong acid.

159. Properties.—Under the ordinary pressure of the atmosphere, carbonic anhydride is a colourless, transparent gas; but, if generated in a confined space, it becomes condensed into a clear liquid, transparent and colourless as water, which, according to Regnault, boils at - 78° C. (- 109° F). At 0° C. (32° F.)—freezing point of water—according to Faraday, it requires a pressure of 38·5 atmospheres to keep it in the liquid state. When a stream of the liquefied body is allowed to escape into the air, it freezes into a snow-white solid.

Carbonic anhydride in the gaseous state is neither inflammable nor a supporter of combustion; in fact, its property of extinguishing a lighted taper is one of the means resorted to for detecting its presence. As, however, some other gases have this property, this alone is not sufficient, and accordingly its property of rendering lime or baryta water turbid or milky is usually the distinctive test for carbonic anhydride. This is caused by its forming a carbonate of the lime or the baryta, which being insoluble is thrown down as a white precipitate; if, however, the carbonic anhydride be continued to be passed into the liquid, the precipitate is re-dissolved, being soluble in excess of *carbonic* acid.

It has a slightly acid taste and smell, and reddens moist blue litmus paper; it is poisonous and destructive to animal life, but is positively essential to the growth of plants. Its specific gravity is about 1½ times that of air, so that it may be collected by downward displacement, and poured from one vessel to another without much loss or admixture. It is also very soluble at 0° C.; water dissolves about 1·8 of its own volume of the gas. It is always present in the air to the extent of about $\frac{1}{2500}$ of the air, 100 pints of air containing ·04 pints of CO_2.

Exp. 68.—In a wide-mouthed bottle put some fragments of

marble or chalk, or any other carbonate; just cover them with water, and add (by means of the safety-funnel) commercial muriatic acid slowly. A brisk effervescence takes place, the gas is given off freely, and, on account of its weight, may be collected by downward displacement (Fig. 41). Several jars having been collected in this way, and their mouths covered with greased glass plates, its properties may be easily examined.

Exp. 69.—Into a jar of the gas insert a lighted taper; the taper is extinguished, and the gas itself does not take fire—that is, it is neither a supporter of combustion, as *oxygen* is, nor is itself combustible, as is hydrogen.

Exp. 70.—Into a jar of the gas pour a little *lime water*, close the jar with a glass plate, and thoroughly shake the two; the lime water will become turbid or milky, and the gas will be found to be absorbed, as the jar may be now inverted without the plate falling off.

Exp. 71.—That carbonic acid is heavier than air, may be proved in a variety of ways. It may be poured from one jar into another like water, its presence or absence being made evident by its action on a burning taper. Or it may be poured into a light jar attached to a balance, and counterpoised by weights in the opposite scale (Fig. 42). Or carbonic anhydride may be raised in a glass bucket from a large jar, and poured into another jar which has been previously tested by a light, and shown to be empty.

Fig. 41.

Exp. 72.—That the precipitate formed by *carbonic anhydride* and lime is re-dissolved in excess of the anhydride, is shown by allowing the delivery-tube of the gas-bottle, generating CO_2, to dip into a beaker or test-glass partially filled with lime water. The clear lime water immediately becomes milky; but, if the gas continue to pass over, after a time (about half an hour) the water becomes quite clear again. If now this clear liquid be boiled, the excess of CO_2 is seen to pass off in bubbles, the liquid becomes turbid, chalk or carbonate of lime (COCao'') being thrown down as an insoluble deposit.

Exp. 73.—The existence of *carbonic anhydride* in the atmosphere may be demonstrated by leaving some lime water exposed to the air in a glass dish; in a very few minutes a white powder will form on the top, which, on agitation, will sink to the bottom. It is insoluble carbonate of lime.

Fig. 42.

Exp. 74.—Its faint acid reaction may be seen by putting a piece of moistened blue litmus paper in a jar of carbonic anhydride, which, after some little time, will be found to be slightly reddened.

The production of carbonic anhydride in combustion has been already proved. That it is also a result of respiration and fermentation, may also be easily proved by experiment.

Exp. 75.—By arranging two bottles, as shown in Fig. 43, and inspiring through the tube A, air will bubble through the lime water in B, before reaching the lungs, but will not cause the lime water to become milky (unless after *long-continued action*); but the air which is expired will bubble through C, and will speedily produce turbidity or milkiness.

Exp. 76.—If a little sugar (brown is best) be dissolved in seven or eight times its weight of warm water, in a flask, and a little dried yeast, previously rubbed down in a mortar with water, be added, fermentation will commence almost immediately, and the carbonic anhydride may be collected in the usual way.

The formation of CO_2 by the burning of charcoal in oxygen is a synthetical experiment, proving that it contains both oxygen and carbon. By the following arrangement of apparatus (Fig. 44), it can be proved to contain nothing but oxygen and carbon, and the proportion in which each enters into its composition can be accurately determined:—

A is a Pepy's gasholder containing oxygen; B, a wash-bottle containing strong sulphuric acid; C a U-tube filled with lumps of calcic chloride; D, a furnace through which passes a hard glass combustion tube, E; in this tube is placed a platinum tray containing very pure carbon. G, a series of bulbs filled with a strong solution of potassic hydrate; H, a U-tube filled with lumps of potassic hydrate.

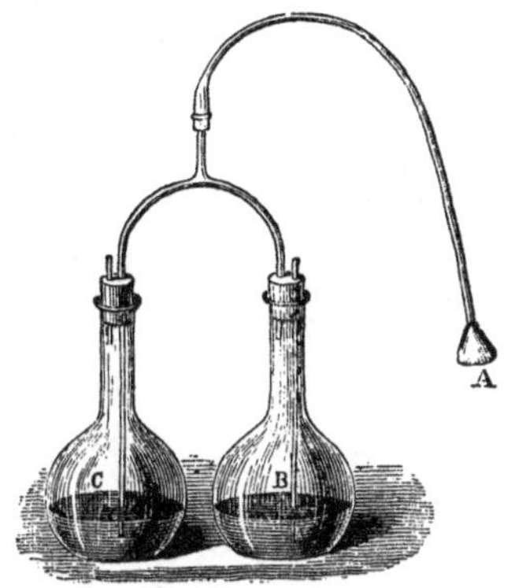

Fig. 43.

The platinum tray with the charcoal, the bulbs G with the potash solution, and the tube H with the potash are each separately weighed as accurately as possible. Oxygen is allowed to stream separately through the whole apparatus, while the charcoal is brought to a red heat in the furnace. The charcoal burns brilliantly in the oxygen, and

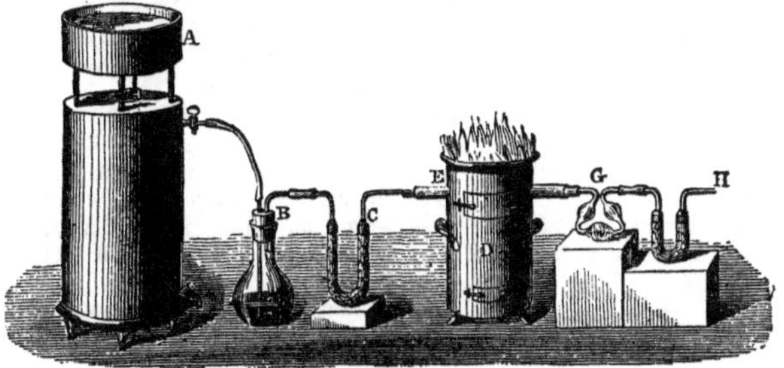

Fig. 44.

the resulting gas is carried forward through the apparatus; in the tube F any possible moisture is arrested, while the potassic hydrate arrests all the *carbonic anhydride*, by combining with it and forming with it *potassic carbonate.* When the apparatus is cold the platinum tray with the charcoal is again weighed, and its loss in weight indicates the amount of *carbon* consumed, while the increase of weight of G and H indicates the amount of *carbonic anhydride* formed; the difference of these two amounts being, of course, the quantity of *oxygen* which has combined with the *carbon.* It is invariably found that the proportions are as 12 to 32—*i. e.*, 12 parts of C require 32 parts of O to produce 44 parts of carbonic anhydride, CO_2.

The *analysis* of *carbonic anhydride* may be effected in the following way :—

Exp. 77.—Cause a stream of the gas to pass through a U-tube, containing *calcic chloride* to dry it, and then through a bulb containing a small piece of *potassium* (Fig. 45). Heat the *potassium* in the

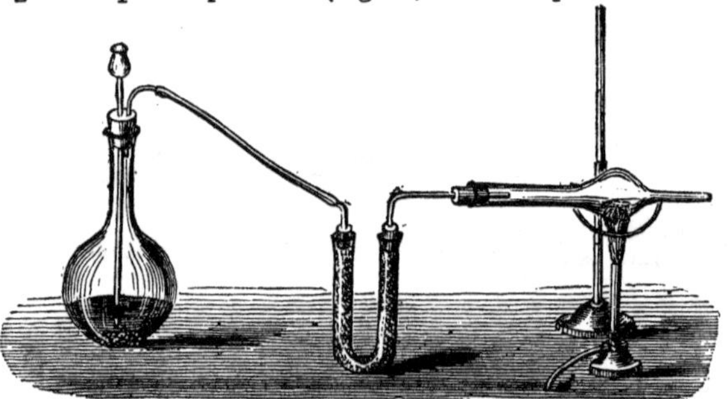

Fig. 45.

gas, the metal will melt and take fire, burning brilliantly at the expense of the *oxygen* contained in the *carbonic anhydride;* when the apparatus is cool, plunge it in water, the alkali is dissolved, and small black particles of carbon are found floating in the liquid.

Exp. 78.—Affix a piece of magnesium ribbon, about 16 centimetres (6¼ inches) in length, to the bowl of the deflagrating spoon, or to a bung, and having set fire to it, plunge it in a jar of *carbonic anhydride;* it will continue to burn, depositing white flakes of magnesic oxide, among which will be interspersed the black particles of *carbon.* To separate these, wash the jar out with a little distilled water, pour it into an evaporating dish, and add a few drops of strong hydrogen chloride, and apply heat.

The magnesic oxide will dissolve, leaving the black particles of *carbon* floating in the clear liquid.

160. *Carbonic anhydride* is usually known in most works on chemistry as *carbonic acid*, and when in union with water it acts as a feeble acid; but the substance *carbonic acid*, $\mathbf{C}OHo_2$ $[CH_2O_3]$, has never been isolated, and is therefore not known. With the alkali metals (K and Na) $\mathbf{C}O_2$ forms two well defined salts, the *carbonate* and *bicarbonate;* but with all its salts, its combination is so weak, that it is separated by effervescence by any of the other acids.

161. Carbonic Oxide.—Symbol, $\mathbf{C}O$. Atomic and molecular weight, 28. Molecular volume, $\square\!\square$. Relative weight, or weight of 1 litre = 14 criths. Specific gravity, ·967.

162. History.—As at present known, *carbon* forms but two compounds with oxygen, of which this is one. It was discovered by Priestley, but its true nature was made known by Cruikshank in 1802.

163. Formation.—It is formed whenever carbon or any carbonaceous substance is burnt with a limited supply of air, as also in the destructive distillation of many substances containing oxygen.

164. Preparation.—1. By passing steam through an iron tube filled with small lumps of charcoal heated to redness, a gas may be collected at the other end, which will be found to burn with a pale blue flame; if you take another jar of this gas, and pour lime water in it, the lime water will be rendered milky.

Three gases are formed in this reaction—*hydrogen, carbonic anhydride*, and *carbonic oxide*—

$$3OH_2 + C_2 = H_6 + \mathbf{C}O_2 + \mathbf{C}O.$$

2. By passing *carbonic anhydride* over charcoal heated to redness, pure carbonic oxide will be formed—

$$\underset{\text{Carbonic anhydride.}}{\mathbf{C}O_2} + C = \underset{\text{Carbonic oxide.}}{2\mathbf{C}O.}$$

3. By passing $\mathbf{C}O_2$ over red-hot iron.

CO may be collected by methods 3 or 4 by means of the gun barrel and furnace used for the liberation of oxygen by chlorine, attaching to one end of the barrel a bottle for the generation of *carbonic anhydride*, and to the other end a leading tube by which the *carbonic oxide* may be collected at the pneumatic trough.

4. By heating *potassic ferrocyanide* (yellow prussiate of potash) with sulphuric acid—

$$Fe''C_6N_6K_4 + 6\mathbf{O}H_2 + 6\mathbf{S}O_2Ho_2 = 6\mathbf{C}O + 2\mathbf{S}O_2Ko_2 + \mathbf{S}O_2Feo'' + 3\mathbf{S}O_2(NH_4O)_2.$$

Potassic ferrocyanide. Water. Sulphuric acid. Carbonic oxide. Potassic sulphate. Ferrous sulphate. Ammonic sulphate.

In this reaction all the carbon appears as *carbonic oxide*, which is remarkably pure.

5. By heating iron or carbon with a carbonate.

6. The most common way of preparing *carbonic oxide* is by heating *oxalic acid* with *sulphuric acid* (by which water is removed from the former)—

$$\begin{cases}\mathbf{C}OHo \\ \mathbf{C}OHo\end{cases} = \mathbf{O}H_2 + \mathbf{C}O_2 + \mathbf{C}O.$$

Oxalic acid. Water. Carbonic anhydride. Carbonic oxide.

The *sulphuric acid* takes no part in the reaction, except removing the water, and thus setting free the two gases. The carbonic anhydride may be removed by causing them to bubble through a wash bottle containing a strong solution of potash; carbonic oxide, not being soluble in that, passes off pure.

165. Properties.—A gas, colourless, but with a faint oppressive odour. It is very poisonous when breathed, more so than CO_2. It is one of the permanently elastic gases, having never been liquefied. It does not support combustion, but burns itself in contact with O with a *pale blue flame*, and producing CO_2; thus—

$$\mathbf{C}O + O = \mathbf{C}O_2.$$

The pale blue flame seen on the top of fires, notably on the surface of coke and charcoal fires, is due to the formation of *carbonic oxide* in the interior of the furnace, and its burning when it comes (in a highly heated condition) in contact with the oxygen of the air at the surface.

CHAPTER XIV.

Nitrogen—Its Preparation and Properties—Compounds of Nitrogen with Oxygen and Hydrogen—Compound of Nitrogen with Hydrogen—Ammonia—Ammonic Salts.

166. Nitrogen.—Symbol, N. Atomic weight = 14. Atomic volume, ☐. Molecular weight, 28. Molecular volume, ☐☐. 1 litre weighs 14 criths. Specific gravity, ·971. Atomicity, ^{v}, which, by mutual saturation of bonds, becomes reduced to $'''$, and to $'$. Evidence of atomicity:—

Nitrogen is monad in *Nitrous oxide*,................$\mathbf{N}N_2$.
,, is triad in *Ammonia*,........................$\mathbf{N}'''H_3$.
,, is pentad in *Ammonic chloride*,..........$\mathbf{N}^{v}H_4Cl$.

167. History.—Discovered by Rutherford, of Edinburgh, in 1772. First observed as a substance in atmosphère which would neither support life nor the burning of a candle, in consequence of which Lavoisier gave to it the name of *azote* (*α, privative; ζωή, life*). Later on, Chaptal give to it the name of *nitrogen,* because he found it forming the active constituent of *nitre.*

168. Occurrence.—Most abundantly in a free state in the atmosphere, of which it forms four-fifths. It has also been found in a free condition in some nebulæ. In combination, it is found in many animal and vegetable bodies, and also as a constituent of many mineral bodies, as the ammoniacal salts, coal-beds, and the nitrates of potash and soda, which are found efflorescing from the soil in India and Peru.

169. Preparation.—1. By the removal of oxygen from

the air, whereby nitrogen is left. This can be best done by the action of phosphorus on the air.

EXP. 78.—If phosphorus be exposed to the air confined in a vessel over water, it will gradually remove the oxygen from the air, forming phosphorous anhydride (P_2O_3), and in two or three days all the oxygen will be removed, leaving nearly pure nitrogen.

EXP. 79.—The same change may be effected in a few minutes if the phosphorus be heated. Place a piece of phosphorus, about half of the size of a nut (previously thoroughly well dried), on a porcelain evaporating dish, and float it in a vessel of water, kindle the phosphorus, and cover it with a jar of air; the phosphorus will burn, and dense white fumes will be formed (fumes of *phosphoric anhydride*, P_2O_5), which will gradually dissolve in the water, and nearly pure *nitrogen* will be found to be left.

2. If air be allowed or caused to pass slowly over ignited copper, the *oxygen* unites with the *copper*, and *nitrogen* is left. The apparatus employed is identical, or nearly so, with that employed for the synthesis of carbonic anhydride:—

A gas-holder, to contain air, and deliver a slow current; a U-tube, filled with calcic chloride, to dry the air as it passes through; an iron or hard glass combustion-tube, filled with granulated copper, and brought to a red heat in a furnace; and a delivery-tube, to deliver at the pneumatic trough.

3. By passing chlorine through an excess of solution of ammonia—

$$\underset{\text{Ammonia.}}{8\mathbf{N}H_3} + 3Cl_2 = \underset{\text{Ammonic chloride.}}{6\mathbf{N}H_4Cl} + \underset{\text{Nitrogen.}}{N_2}.$$

There is some little danger in this experiment, as unless a *great deal of care is taken*, and a *very large* excess of *ammonia* is present, nitrous chloride (chloride of nitrogen) will be formed, which will almost as certainly explode with dangerous violence.

4. By heating *ammonic nitrite*, which entirely breaks up into *water* and *nitrogen*—

$$\underset{\text{Ammonic nitrite.}}{\mathbf{N}'''O(N^vH_4O)} = \underset{\text{Water.}}{2\mathbf{O}H_2} + N_2.$$

$$[NH_3HNO_2 = 2H_2O + N_2.]$$

5. By heating a mixture of *ammonic chloride,* with *potassic* or *sodic nitrite*—

$$\mathbf{N}H_4Cl + \mathbf{N}ONao = NaCl + 2\mathbf{O}H_2 + N_2.$$

Ammonic chloride. Sodic nitrite. Sodic chloride. Water.

170. Properties.—*Nitrogen* is one of the permanently elastic gases, having never been liquefied by cold or pressure; it has neither colour, taste, nor smell. It has no action on litmus or turmeric paper, nor does it whiten lime-water. It is not absorbable by potash. It is not a combustible, nor does it support combustion. It is so slightly soluble in water, that it may be almost regarded as insoluble—water, which has been boiled, dissolving it to the extent of 1½ to 2 per cent. Although by itself it is fatal to animal life, it is not poisonous; it kills by depriving the animal of the *oxygen* necessary to existence. It has scarcely any tendency to unite with any of the elements except *boron, titanium,* and a few of the rarer metals; when it does enter into combination, it does so with great difficulty, and its compounds generally are characterized by their instability, and their liability to decomposition, frequently with explosive violence.

The striking contrast between *nitrogen* and *oxygen* will be seen from this list of the purely negative properties of the former, and yet, in combination, *nitrogen* forms part of the most powerful and active substances known, as for instance, *nitric acid* (aquafortis), and *ammonia* (hartshorn), the extremes of acidity and alkalinity. It forms part of the strongest vegetable poisons, as *strychnia, morphia,* and *prussic acid;* and it is also an essential component of all the more important and more valuable forms of food, as bread, milk, and animal flesh.

The compounds of *nitrogen* are called *nitrides,* the most important being those which it forms with *oxygen, hydrogen,* and *hydroxyl.*

The weight of a litre of dry air, free from *carbonic anhydride,* at 0° C., and with the barometer standing 760 m. m., has been found to be 1·2932 grammes.

Besides these four—O, N, CO_2, and OH_2—which are always present, and may be considered essential to the very constitution of the atmosphere, there are found variable traces of *nitric acid*, NO_2Ho [HNO_3], and *ammonia*, NH_3; but, in such very minute proportions, that they can with difficulty be detected, even when very large quantities of air are examined. They exist in greater abundance after a severe thunderstorm, electricity having the chief part in causing their elements to unite; and, in the rain which generally accompanies a severe storm, they may frequently be found. The rain washes them out of the air, and brings them down to the soil. Minute as their proportions are, they are of the utmost use (if they are not absolutely essential) to the health of the vegetable world.

Near large towns, especially manufacturing towns such as Leeds, small quantities of other gases, as *sulphurous anhydride*, SO_2, from the burning of coal, and *hydric sulphide*, SH_2, from the decomposition of animal matter and various manufacturing processes, have been detected.

Miller gives the following as the average composition of a *litre* (1,000 c. c.) of air:—

	Cubic Centimetres.
Oxygen,	206·1
Nitrogen,	779·5
Aqueous vapour (about),	14
Carbonic anhydride,	0·4
Nitric acid, Ammonia, Carburetted hydrogen,	traces.
	1000·0

171. Compounds of Nitrogen with Oxygen and Hydroxyl.—As we said before in Chapter III., page 29, when illustrating the *Law of Multiple Proportions*, the compounds of *nitrogen* and *oxygen* are not only very important in a chemical point of view, but they are very interesting and instructive as being the most regular series we know of, the nitrogen remaining constant, while the oxygen regularly increases—

NAME.	ATOMIC FORMULÆ.		GRAPHIC FORMULÆ.
Nitrous oxide,............	ON_2	$[N_2O]$	
Nitric oxide, ...	$\left\{\begin{matrix}NO\\NO\end{matrix}\right.$	$[N_2O_2]$	
Nitrous anhydride, ...	$\left\{\begin{matrix}NO\\O\\NO\end{matrix}\right.$	$[N_2O_3]$	
Nitric peroxide, ...	$\left\{\begin{matrix}NO_2\\NO_2\end{matrix}\right.$	$[N_2O_4]$	
Nitric anhydride, ...	$\left\{\begin{matrix}NO_2\\O\\NO_2\end{matrix}\right.$	$[N_2O_5]$	
Nitrous acid, ...	NOHo	$[HNO_2]$	
Nitric acid,...............	NO_2Ho	$[HNO_3]$	

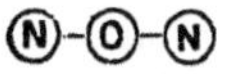

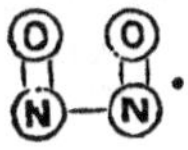

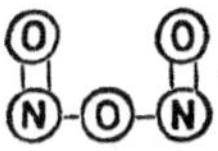

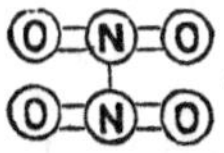

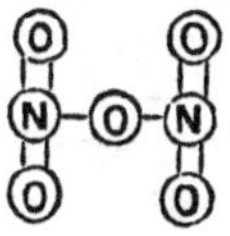

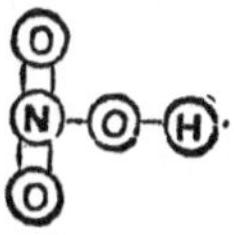

The above table is taken from Professor Frankland's *Lecture Notes for Chemical Students*, page 61.

172. Nitric Acid.—Called also *hydric nitrate* and *aquafortis*. Symbol, NO_2Ho $[HNO_3]$. Molecular weight, 63. Molecular volume, ☐☐ . One litre of nitric acid vapour weighs 31·5 criths. Specific gravity of liquid = 1·52. Boiling point, 84·5° C. (184° F.) Fuses at −50° C. (−58° F.)

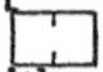

Although the attraction between oxygen and nitrogen is exceedingly feeble, this acid is one of the strongest and most important known—formerly (and still commonly) called *aquafortis* (strong water), from its power of dissolving nearly all the metals. Its composition was first ascertained by Cavendish in 1785.

173. Preparation.—1. By the passage of electricity through a mixture of *oxygen* and *nitrogen* in the presence of moisture; consequently, *nitric acid* is formed always in small quantities by lightning. In combination with *ammonia*, $\mathbf{N}H_3$, as ammonic nitrate, it may nearly always be detected in rain water.

2. Whenever organic matter, containing nitrogen (chiefly animal), is slowly oxidized in the presence of powerful bases (as *soda*, *potash*, or *lime*), a nitrate is formed. Whenever a nitrate is found in well or spring water, it is a proof of the existence of decaying animal matter in the soil through which it has passed.

This plan of obtaining nitrates for the manufacture of gunpowder—viz., by the oxidation of animal matter, has been largely adopted by France, Sweden, and some other countries.

174. Manufacture.—*Potassic nitrate* is found as an efflorescence on the soil in tropical climates, especially in some districts in India; and in Chili, *sodic nitrate* is found in a similar condition.

In chemical works and laboratories nitric acid is always made by distilling one of these salts with *concentrated sulphuric acid*—

$$\mathbf{N}O_2Ko + \mathbf{S}^{vi}O_2Ho_2 = \mathbf{S}^{vi}O_2HoKo + \mathbf{N}^{v}O_2Ho.$$

Potassic nitrate. Sulphuric acid. Hydric potassic sulphate. Nitric acid.

$$[KNO_3 + H_2SO_4 = HKSO_4 + HNO_3.]$$

In practice, the manufacturer uses iron cylinders or retorts, which are placed in a furnace, and connected by a series of pipes with stoneware bottles. He can thus apply a greater heat than the glass vessels of the chemist would stand, and can obtain his *nitric acid* with half the quantity of sulphuric acid. The reaction takes place in two stages, the heat required for the last stage being very great.

(1.) $2\mathbf{N}^{v}O_2Ko + \mathbf{S}O_2Ho_2 = \mathbf{S}O_2HoKo + \mathbf{N}O_2Ko + \mathbf{N}O_2Ho.$

Potassic nitrate. Sulphuric acid. Hydric potassic. sulphate. Potassic. nitrate. Nitric. acid.

(2.) $\mathbf{S}O_2HoKo$ +	$\mathbf{N}O_2Ko$ =	$\mathbf{N}O_2Ko_2$ +	$\mathbf{N}O_2Ho$.
Hydric potassic sulphate.	Potassic nitrate.	Potassic sulphate.	Nitric acid.

The explanation of this is as follows :—*Sulphuric acid* forms with *sodium* or *potassium* two salts, a *hydric* or acid *sulphate*, and a *neutral sulphate*. The *acid* or *hydric sulphate* is very soluble and readily fusible, and can therefore be extracted from the glass retort without risk of breakage; while the *neutral sulphate* is less soluble, and positively infusible in glass vessels.

175. Properties.—*Nitric acid* when pure is perfectly colourless, and when strong, fumes in the air; it is however generally slightly yellow (always so, when made by the manufacturer's process) from the presence of some of the lower oxides of *nitrogen* which are dissolved in it.

It is one of the most powerful acids, dissolving all the metals except gold and platinum, and forming with their oxides or hydrates a class of salts called *nitrates*, all of which are soluble in water—

$\mathbf{O}KH$ +	$\mathbf{N}O_2Ho$ =	$\mathbf{N}O_2Ko$ +	$\mathbf{O}H_2$.
Potassic hydrate.	Nitric acid.	Potassic nitrate.	Water.

$\mathbf{Pb}O$ +	$2\mathbf{N}O_2Ho$ =	$\left\{\begin{matrix}\mathbf{N}O_2\\ Pbo''\\ \mathbf{N}O_2\end{matrix}\right.$ +	$\mathbf{O}H_2$.
Plumbic oxide.	Nitric acid.	Plumbic nitrate.	Water.

It is an intensely corrosive liquid, acting on organic substances with great violence. It destroys nearly all vegetable colours, and stains the skin a permanent yellow; great care is requisite in experimenting with this acid. On account of its decomposing, and parting with its *oxygen* so easily, it is a powerful oxidizing agent.

176. Tests.—The following are the laboratory tests for nitric acid.

1. It should thoroughly bleach a dilute solution of sulphate of indigo. 2. It dissolves *copper*, forming a blue solution, and giving off fumes which turn red in contact with air. 3. Poured on *morphia*, it turns it of a brilliant red colour. 4. Poured on *proto-*

sulphate of iron, it turns it of an olive-brown colour by oxidation. 5. It should be colourless, and leave no residue on evaporation. 6. When freely diluted with water, it should give no precipitate with *nitric acid* or *chloride of barium*.

177. Nitric Anhydride.—Symbol, $\mathbf{N}_2O_5$. Molecular weight, 108. Probable molecular volume, □□. Fuses at 29·5° C. (85° F.), and boils at 45° C. (113° F.)

First produced as a white, crystalline, unstable solid in 1844.

178. Nitrous Oxide, or **Nitrogen Protoxide**—Called also **Laughing Gas.**—Symbol, $\mathbf{O}N'_2$ [N_2O]. Atomic and molecular weight = 44. Molecular volume, □□. 1 litre weighs 22 criths. Specific gravity = 22. Fuses at − 101° C. (149·8° F.). Boils at − 88° C. (126·4° F.)

179. Preparation.—By heating *ammonic nitrate*, $\mathbf{N}^vO_2(N^vH_4O)$ [H_4N,NO_3], in a retort or flask, it is decomposed into water and nitrous oxide gas—

$\mathbf{N}^vO_2(N^vH_4O)$	=	$2\mathbf{O}H_2$	+	$\mathbf{O}N_2$.
Ammonic nitrate.		Water.		Nitrous oxide.
[H_4N,NO_3	=	$2HO_2$	+	N_2O.]

180. Properties.—*Nitrous oxide* is a gas, colourless, transparent, with a faint sweet taste and smell. It does not burn itself, but it supports combustion *almost as well as oxygen*. This property is due to the ease with which *nitrous oxide* breaks up into its constituents.

It will not support life, but can be breathed for a time safely with impunity, although it has a peculiar effect on the brain and nerves. If mixed with air, and breathed, it occasions a species of intoxication of a very peculiar kind, which generally manifests itself by uncontrollable bursts of laughter; hence its name of *laughing gas*. Sometimes, however, it produces the opposite effect, viz., that of crying; neither effect is, however, of a lasting nature. If breathed in a perfectly pure state, it produces transient insensibility; and in this condition is used largely in surgical operations, especially in dentistry, its use not being attended with the same danger as that of *chloroform*.

At 0° C. (32° F.) it liquefies under a pressure of 30 atmospheres, and at – 88° C. (126·4° F.) under that of one atmosphere; while at a temperature of about – 101 C. (449·8° F.) it freezes into a transparent crystalline solid.

That it is a true chemical compound, and not a mechanical mixture of N and O, as air is, may be shown by the two tests for showing the pressure of free oxygen —viz., the nitric oxide test, which, with free O, forms deep orange-red fumes of *nitrous anhydride* soluble in water, and by the *potassic pyrogallate* test, which absorbs free oxygen. With *nitrous oxide,* however, neither of these tests give any reaction.

Nitrous oxide is slightly soluble in water. At 15·5° C. (60° F.), one volume of water dissolves about three-fourths of its volume of gas, or at 0° C. (32° F.) one volume of warm water dissolves 1·3 volumes of this gas; in warm water it is less soluble.

Exp. 80.—Half fill a test tube with this gas over water, close the tube firm by the thumb under water, and then raising the tube (keeping it firmly closed), agitate briskly the gas and water together. On now opening the tube under water a considerable rush upwards of water into the tube will take place, showing the gas to have been dissolved in the water. By these means *nitrous oxide* may be distinguished from *oxygen.*

181. Nitric Oxide.—Symbol, $\left\{\begin{matrix}\mathbf{N}O\\ \mathbf{N}O\end{matrix}\right.$, or $'\mathbf{N}''_2O_2$[NO]. Molecular and combining proportion = 60.* Molecular volume anomalous, ⊞. 1 litre weighs 15 criths.

* Miller, Bloxam, and other chemists give the molecular weight of *nitric oxide* as 30; this arises from considering the molecule to be NO ◫; but according to Frankland's theory of atomicity, NO could not have a separate existence, there being bonds unsatisfied; it is therefore a compound radical (see pages 60, 61), and as such can enter into combination, but cannot exist alone. The true molecule, therefore, of *nitric oxide* must be as given above, $\left\{\begin{matrix}\mathbf{N}O\\ \mathbf{N}O\end{matrix}\right.$, or $'N''_2O_2$, and its molecular weight must therefore be 60.

Specific gravity, 15 in the hydrogen scale; or 1·039, air being taken as unity.

182. Preparation.—Dilute *nitric acid* with water until it attains a specific gravity of 1·2, then pour it on some copper turnings or clippings contained in a retort or flask with leading tube; the retort immediately becomes filled with deep orange-red fumes, and a colourless gas may be collected at the pneumatic trough over water—

$$3Cu + 8\mathbf{N}^{v}O_2Ho = 3Cu\begin{matrix}\mathbf{N}O_2 \\ o \\ \mathbf{N}O_2\end{matrix} + 4\mathbf{O}H_2 + \mathbf{N}''_2O_2.$$

Copper. Nitric acid. Cupric nitrate. Water. Nitric oxide.

$$[3Cu + 8HNO_3 = 3(Cu,2NO_3) + 4H_2O + 2NO.]$$

Mercury may be substituted for copper with the same effect.

183. Properties.—A transparent, colourless gas, with a strong, disagreeable, suffocating odour; does not burn itself, nor support combustion in an ordinary way, but if any combustible body, such as phosphorus, charcoal, sulphur, &c., *when burning vigorously*, be plunged into the gas, it will deflagrate with great brilliancy. This arises from the heat of the burning body decomposing the *nitric oxide;* the combustible then unites with the *oxygen*, setting free the *nitrogen.*

Nitric oxide has never yet been liquefied, and is therefore regarded as one of the permanently elastic gases; it is all but insoluble, water dissolving only $\frac{1}{20}$ of its volume of this gas.

In contact with *free oxygen*, it instantaneously forms deep red fumes of *nitrous anhydride* and *nitric peroxide*, which are soluble in water; on this account it becomes a very valuable test for *oxygen*, and affords means for detecting even the smallest admixture of *oxygen* with other gases.

In the manufacture of *sulphuric acid*, it plays (as we shall see) a most important part, alternately absorbing *oxygen* from the air, and yielding it up to the *sulphuric acid*, which is forming.

184. *Nitric oxide* is absorbed by a solution of ferrous sulphate, which becomes of a deep reddish-brown colour; it is also dissolved by strong *nitric acid*, which turns first yellow, then brown, and ultimately green.

Exp. 81. Fill a small gas jar with water, coloured blue with litmus, and pass into sufficient nitric oxide to fill about one-third of the jar, pass into it a few bubbles of oxygen, deep red fumes are formed which are quickly dissolved, and the blue colour of the solution will be changed to red. If the *nitric oxide* and *oxygen* be perfectly pure, and properly added, it is possible to completely absorb both gases.

185. Nitrous Anhydride.—Symbol, $\mathbf{N}'''_2O_3$. Probable molecular weight, 76. Probable molecular volume, $\square\square$.

186. Preparation.—1. By mixing 4 vols. of perfectly dry *nitric oxide* with 1 of dry *oxygen*, brownish-red fumes of *nitrous anhydride* are formed. These at $-18°$ C. (about 0° F.) are condensed to a blue liquid.

2. By heating *nitric acid* (specific gravity, 1·25) with starch.

3. By dissolving silver in cold nitric acid.

187. Properties.—A blue, volatile, unstable liquid, which boils at $-18°$ C. (0° F.), forming red vapours. If water be added to it, it is converted into *nitrous acid*, but on the addition of more water is decomposed into *nitric acid* and *nitric oxide*.

There is a great deal of doubt as to the real nature of this compound; it may be, and sometimes is, regarded as a compound of N_2O_5 with NO; thus—

$$3\mathbf{N}_2O_2 \quad = \quad N_2O_2 \quad + \quad 4NO.$$

188. Nitrous Acid (*Hydric Nitrite*).—Symbol, $\mathbf{N}OH$. Molecular weight = 47.

189. Preparation.—By mixing liquefied *nitrous anhydride* with a small quantity of water—

$$\underset{\text{Nitrous anhydride.}}{N_2O_3} \quad + \quad OH_2 \quad = \quad \underset{\text{Nitrous acid.}}{2\mathbf{N}OHo.}$$

190. *Nitrous acid*, in presence of an excess of water, is decomposed into *nitric acid* and *nitric oxide*, as mentioned above.

Nitrous acid is capable of absorbing *oxygen*, becoming converted into *nitric acid*, when it acts as a *reducing* agent; or of parting with *oxygen* and *water*, and being changed into *nitric oxide*, when it is an *oxidizing* agent.

With *metallic oxides* and *hydrates*, it forms a class of salts, called *nitrites*, which are, however, of very little practical importance.

191. Nitric Peroxide.—Symbol, $N^{iv}_2O_4$. Molecular weight = 46 to 92. Molecular volume below 0° C., ☐ ; at 100° C., ☐☐ . One litre weighs 23 to 46 criths.

192. Preparation.—1. By the union of *nitric oxide* and *oxygen*—

$$N''_2O_2 \quad + \quad O_2 \quad = \quad N^{iv}_2O_4.$$

2. The most convenient method of obtaining *nitric peroxide* is by heating nitrate of lead in a retort or glass tube. Deep red fumes are given off, which consist of *nitric peroxide* and *oxygen*.

$$N_2O_4Pbo'' \quad = \quad 2Pb''O \quad + \quad 2N_2O_4 \quad + \quad O_2.$$

These fumes at a temperature of 22° C. (71·6° F.) may be condensed to a red liquid; and, if it be free from water, at 12° C. (10·4° F.), may be obtained in yellow prismatic crystals. A very small trace of water is enough to decompose it, *nitric acid* being formed, and nitric oxide set free.

The exact constitution of *nitric peroxide*, like that of nitrous anhydride, is somewhat uncertain.

With metallic oxides and hydrates it produces a mixture of *nitrites* and *nitrates*.

CHAPTER XV.

Compounds of Nitrogen with Hydrogen—Ammonia—its History, Preparation, and Properties—Ammonium—Ammonia Salts.

Although it is suspected that *hydrogen* and *nitrogen* unite in several proportions in an indirect manner, only one has been yet successfully isolated, the well known *ammonia* or *hartshorn*.

193. Ammonia.—Symbol, NH_3 [H_3N]. Atomic and molecular weight = 17. Molecular volume □□ . One litre weighs 8·5 criths. Relative weight, 8·5. Specific gravity, 0·59. Fuses at − 75 C. (−103 F.); boils at −38·5° C. (−33·7° F.)

194. History.—First obtained by the Arabs from camel's dung, obtained near the temple of Jupiter Ammon, in Lybia, North Africa, and hence called *sal-ammoniac.*

195. Occurrence.—In the atmosphere, in very minute quantities, and formed whenever animal or vegetable matter containing *nitrogen* decomposes, in the presence of moisture, or is distilled at a red heat. It is consequently present in all manures, especially in guano, which is the excrement of sea birds, and in the urine of all animals. It is most important in the nourishment of plants, and in this way may be regarded as the most necessary constituent of manure. Its presence in animal matter, and liberation on destructive distillation, may be shown by the following experiment.

Its principal source of supply is from the waste liquors collected during the distillation of coal in the manufacture of gas: so that we may regard its sources of manufacture as two:—

1. By the destructive distillation of horn, bone, or any other animal substance, and also some very few vegetable substances which contain *nitrogen.*

2. In the process of making coal, ammonia is set free from the compounds of nitrogen which exist to a greater or lesser extent in all coal, and which (NH_3) having to be purified, is retained in the waste liquors, from whence it is obtained by the action of acids, either as a *sulphate* or *chloride of ammonia.*

196. Preparation.—By heating a mixture of lime and *ammonic chloride (sal-ammoniac)*—

$2NH_4Cl$	+	$CaCO$	=	$CaCl_2$	+	$2NH_3$	+	OH_2.
Ammonic chloride.		Lime.		Calcic chloride.		Ammonia.		Water.

In doing this, the *ammonic chloride* and the *lime* must both be *perfectly* dry, and thoroughly powdered, or the lime should be freshly slaked, and allowed to cool.

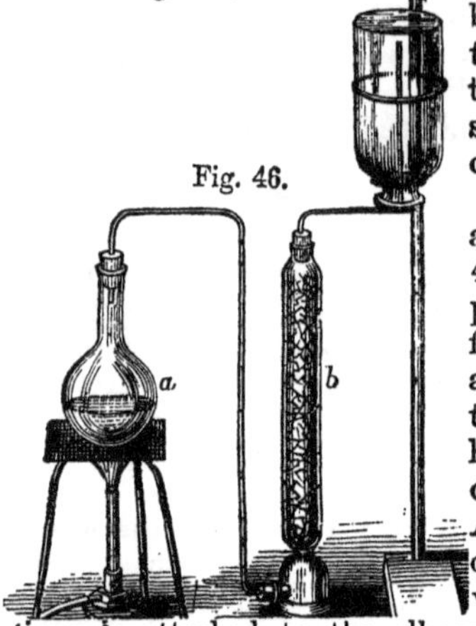

Fig. 46.

Such an arrangement as that shown at Fig. 46, is suited for the purpose — an ordinary flask containing the lime and ammonic chloride. To this is fitted an elbow tube, having each arm about 8 or 10 c. m. (3 or 4 inches). A drying tube (chloride of calcium tube), filled with small lumps of quick-lime, is attached to the elbow-tube; and to the other extremity of the drying tube is attached another elbow-tube, having one long arm, that it may reach to the top of the bottle, which is to act as a gas-receiver. The gas, being much lighter than air, can be collected by upward displacement. As gaseous ammonia is very irritating, it will be advisable to cover the mouth of the bottle with a piece of cardboard, through which the delivery tube can pass.

In order to know when the bottle is full, hold a piece of moistened yellow turmeric paper near to and just above its mouth. If the gas is overflowing, it will rise and turn the turmeric paper brown.

This is the common method employed in preparing *ammonia* for experiment.

197. Properties.—*Ammonia* is a gas, colourless and transparent; it possesses a very pungent odour, and acts powerfully on the eyes; it has an acrid, biting taste; its power of stimulating all the nerves is very great; and it is on this account largely used as smelling salts to check feelings of faintness; its solution is also used medicinally as a stimulant. It is powerfully alkaline, and extremely soluble. Water at 0° C. dissolves 1,050 times its volume of the gas, forming a liquid of specific gravity ·853, at 15·5° C. (60° F.) Water dissolves 727 times its volume, the liquid formed having a specific gravity of ·886.

Its alkalinity and its solubility may both be shown by the following experiment:—

EXP. 82.—**Fill a glass basin, or small glass pneumatic trough, with water coloured with litmus, and just feebly reddened by the addition of any acid. Remove the jar of *ammonia* carefully with its mouth well closed, plunge it, mouth downwards, in the solution of litmus and unclose it, the liquid will rush rapidly upwards in the flask, the ammonia will be absorbed, and the red liquid will become blue.**

Gaseous ammonia becomes liquid under ordinary pressure at $-40°$ C. ($-40°$ F.), and a transparent solid at $-75°$ C. ($-103°$ F.) Under a pressure of 7 atmospheres it liquefies at 15·5° C. (60° F.) Good boxwood charcoal will absorb in the cold about 90 times its bulk of *ammonia.*

Ammonia, when pure, is fatal to life, producing great irritation of the lungs and spasm of the glottis. It will not support combustion; and on account of its own igniting point being very high, it does not burn by any means freely itself; but in a jar of *oxygen* or mixed with oxygen it burns with a pale green flame.

Ammonia neutralizes the strongest acids, and forms with them an important series of salts, known as the ammonium salts. In all these the atomicity of the nitrogen is ^{v}:—

$$\mathbf{N}'''H_3 + HCl = \mathbf{N}^{v}H_4Cl$$

Ammonia. + Hydrochloric acid. = Ammonic. chloride.

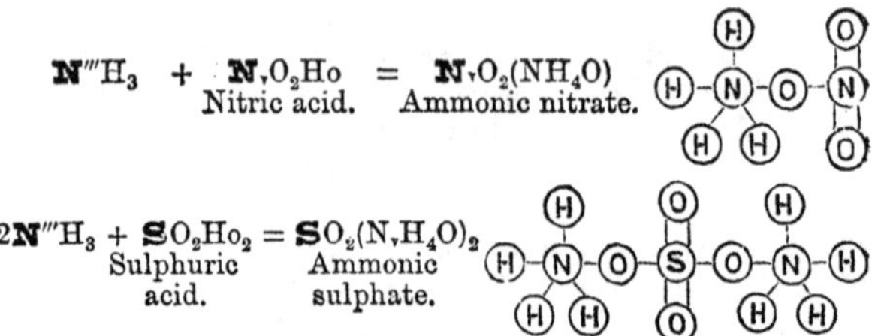

Any volatile acid brought into an atmosphere containing ammonia, produces a white cloud, formed by the union of the *ammonia* and the *acid*, and the formation of a salt of ammonia (all of which are white), in a very minute state of sub-division. This property is usually employed as a test to detect the presence of very small quantities of ammonia.

Exp. 83.—Mix a very little *hydrochloric acid* with about half its bulk of water, dip the end of a glass rod in the mixture, and hold it near to the mouth of a flask of *ammonia*, dense white fumes will appear around the rod.

198. Synthesis of Ammonia.—The synthetical formation of *ammonia* is strikingly shown by the following experiments:—

Make a mixture of *iron filings* and *caustic potassa*, in about the proportion of 15 to 1, and call it mixture No. 1. Next make one of *iron filings* and *potassic nitrate*, in the same proportions, and call it mixture No. 2.

Exp. 84.—Put some of mixture No. 1 in a test tube, and heat in a Bunsen flame, a gas will be given off, which may be collected over water at the pneumatic trough, and which on examination will be found to be *hydrogen*. Iron at a high temperature displaces *hydrogen* from *caustic potash*.

Exp. 85.—Heat in same way some of mixture No. 2. A gas will be given off, which may be collected over water, and which on examination will prove to be *nitrogen*.

Exp. 86.—Put equal quantities of each mixture Nos. 1 and 2, well incorporated together, in a test tube, and apply heat. A gas will now be given off, which cannot be collected over water, but may be over *mercury*, and which by all the tests can be proved to be *ammonia*.

The *hydrogen* and the *nitrogen*, at the moment of being set free, seize on each other, combine together, and form *ammonia.*

Elements at the moment they are set free from their combinations are said to be in the *nascent* or new-born condition, and at that instant their powers of combination are very highly exalted.

199. Analysis of Ammonia.—The constitution of *ammonia,* that it is composed of one measure of *nitrogen* and three measures of *hydrogen,* which in the act of combining become condensed to two measures, thus—

$$\boxed{N} + \boxed{H \mid H \mid H} = \boxed{H_3 \mid N}$$

may be shown by analyzing it by means of electricity.

Ammonia can also be decomposed by *chlorine,* by passing *chlorine* through an excess of solution of *ammonia,* as mentioned in Chapter XIV., page 147. For every 3 volumes of *chlorine* employed, 1 volume of *nitrogen* is left; and, as 3 of *chlorine* take up 3 of *hydrogen* to form 3 of HCl, the composition of *ammonia* must be $\mathbf{N}H_3$. The HCl so formed combines with the excess of ammonia to form the solid salt ammonic chloride, $\mathbf{N}H_4Cl$.

200. Ammonium.—Symbol, $\mathbf{N}^{v}H_4$. Molecular symbol, $\left\{\begin{matrix}\mathbf{N}H_4\\ \mathbf{N}H_4\end{matrix}\right.$. The monad radical NH_4 has never yet been obtained in a free state; but it can be obtained combined with Hg, as an amalgam—a property usually considered as confined entirely to metals. It combines with the acids, and forms a large series of *ammonic salts,* perfectly analogous in crystalline form and other properties to the salts of *potassium.* On these two grounds, chemists usually consider the radical as a hypothetical metal *ammonium,* NH_4, usually written *Am,* and forming *ammonoxyl, Amo,* analogous to *Ho, Ko, Nao,* &c.

The compounds of *mercury* with metals all possess metallic lustre, and are called *amalgams.* It may be prepared in two ways—

1. By the electrolysis of ammonic chloride, the negative electrode being *mercury*, and the positive a plate of *platinum*, the mercury swells up, and a spongy metallic mass is formed—*ammonic amalgam.*

2. Prepare an amalgam of sodium or potassium, and pour it into a solution of ammonic chloride, slightly warmed—the mercury swells up enormously, an amalgam being formed; while sodic or potassic chloride is formed simultaneously—

Hg_nNa_m	+	$m\mathbf{N}H_4Cl$	=	$Hg_n(N^vH_4)m$	+	$mNaCl.$
Sodic amalgam.		Ammonic chloride.		Ammonic amalgam.		Sodic chloride.

Exp. 87.—Dissolve a piece of *sodium* about the size of a pea, in about 2 c. c. of mercury in a test tube, or rub them down together in a glass mortar, the two metals unite together suddenly with flame, and with a slight explosion. When cold, pour the amalgam into a watch-glass, and cover with a saturated solution of *ammonic chloride,* the amalgam will gradually swell up and become pasty, and will even sometimes float on water.

The *ammonium* cannot be obtained in a separate form from this amalgam, as, if decomposed by heat, or thrown into water, *hydrogen* is disengaged, *ammonia* is given off or dissolved, and pure mercury is left—

$2Hg_n(N^vH_4)m$	=	$2nHg$	+	$2m\mathbf{N}H_3$	+	$mH_2.$
Ammonic amalgam.		Mercury.		Ammonia.		

201. Ammonic Salts.—*Ammonium* forms salts with all the acids, in analogy with the salts formed by *potassium* and *sodium.* Nearly all these salts are volatile, a property which very much adds to their value in the laboratory.

One of the most important compounds of *ammonium* for laboratory use is *ammonic sulphide,* which is obtained thus—

Exp. 88.—Measure out two equal portions of a solution of *ammonia,* or *ammonic hydrate* (ammonia). Pass through one of these a brisk current of sulphuretted hydrogen gas, as long as it continues to be absorbed. (The apparatus used for the production of chlorine water will be found particularly suitable.) Then add the second portion to it, and a solution of *ammonic* sulphide $\mathbf{N}^vH_4(HS'')'$ is obtained.

When a solution of *ammonia* is *completely* saturated with sulphuretted *hydrogen*, a solution of *hydric-ammonic sulphide* is obtained, in which the compound radical *hydrosulphyl* has taken the place of the *hydroxyl*—

AmHo*	+	$\mathbf{S}H_2$	=	AmHs	+	$\mathbf{O}H_2$.
Ammonic hydrate.		Sulphuretted hydrogen.		Hydric-ammonic sulphide.		Water.

On then adding AmHo to the solution, *ammonic sulphide* is obtained. Thus—

AmHs	+	AmHo	=	$\mathbf{S}Am_2$	+	$\mathbf{O}H_2$.
Hydric-ammonic sulphide.		Ammonic hydrate.		Ammonic sulphide.		Water.

CHAPTER XVI.

Sulphur—History—Occurrence—Preparation—Properties—Allotropic Modifications—Uses—Compounds of Sulphur with Positive Elements—Sulphuretted Hydrogen—Occurrence—Preparation—Properties—Reactions—Uses—Hydrosulphyl—Carbonic Disulphide—Properties—Reactions.

202. Sulphur.—Symbol, S or S_2. Atomic weight = 32. Molecular weight, 64. Molecular volume, [figure: volume box], at 1,000° C. (1,768° F.) Below 815° C. (1,500° F.), volume only ⅓. One litre of sulphur vapour weighs 32 criths. Melting point, 115° C. (239° F.) Boiling point, 446° C. (836° F.) Atomicity, ″, ^iv^, and ^vi^.

203. History.—*Sulphur* has been known and used as a bleaching agent from very early times, but its properties were first thoroughly examined by Priestley in 1774.

* The *hydrate of ammonia* is purely hypothetical, since no chemical combination takes place between *water* and *ammonia;* but still, in combination, it behaves, as shown in the above equation, as though there were a definite hydrate formed.

204. Occurrence.—*Sulphur* is most abundantly distributed throughout all nature, either free or in combination with other elements. Most of the sulphur used in England is found in a native or uncombined state, in beds of blue clay in Sicily, near Mount Etna. It is also found as a sublimation around the mouths of volcanoes, where it is mixed with, and condensed on the sands and gravel; it is in this condition it is found around the solfataras of Guadaloupe, Pouzzales, &c. It is also found distributed through nature in combination with the metals, as *iron pyrites (ferric sulphide)*, FeS_2; with *copper* as *copper pyrites (cuprous sulphide)*, $(\mathbf{FeCu})S_2''$; with *lead* as *galena (plumbic sulphide)*, PbS; and with *zinc* as *blende (zincic sulphide)*, ZnS. It also occurs largely in the form of *sulphates*, combined with the following metals, Ca, Mg, Ba, Sr, Na, &c. It is also found in many vegetables, especially in cabbage, and nearly all of the *cruciferæ*. It also enters into the muscular tissue of animals, and into the white of egg.

205. Extraction.—When it is mixed with much earthy impurities, as in the gravels of Sicily, &c., it is roughly distilled in earthenware retorts, the sulphur being condensed in earthenware receivers. This does not remove all impurities, so that it undergoes a second distillation in iron retorts, the vapour being carried into large brickwork chambers, on the walls of which it is condensed in the form of flowers of sulphur. It is collected in that state, and the greater portion melted and run off into wooden moulds to form the roll sulphur or brimstone of commerce.

It is obtained from *iron* and *copper pyrites* by either distilling them or roasting them in a closed kiln, according to the purpose for which the *sulphur* is to be used; but the sulphur so obtained is nearly always impure, being associated with arsenic and other bodies.

206. Properties.—Sulphur is a yellow, brittle solid, a bad conductor of heat and electricity, tasteless, and with a slight odour. It is absolutely insoluble in water, but is slightly soluble in boiling anhydrous alcohol, and in some of its forms is soluble in hot oil of turpentine and in carbonic disulphide, $\mathbf{C}S_2''$.

It is particularly inflammable, burning with a pale blue flame, and producing suffocating odours of sulphurous anhydride, $\mathbf{S}O_2$.

The most remarkable thing with reference to sulphur is its curious deportment with reference to heat. At 115·5° C. (240° F.) it melts, and becomes a transparent, limpid, pale yellow liquid; if the heat be increased, at 138° C. (280° F.) it turns brown, and at about 176° C. (350° F.) it becomes almost black and very thick; on continuing the heat it becomes more and more viscid and treacly, until at about 250° C. (482° F.) it is perfectly opaque, and cannot, on account of its viscidity, be poured into another vessel. At 315° C. (600° F.) it becomes again liquid, and if now poured into cold water it becomes an elastic and plastic solid, varying in colour from pale amber to deep brown. In this condition, in which not one of its properties seem to resemble those of ordinary sulphur, it is largely used for taking moulds for electrotyping purposes. In a few days at most, sometimes in a few hours, and suddenly at a temperature of 100° C., it returns to the hard brittle condition.

207. Allotropic Modifications.—*Sulphur* is one of the most important elements in establishing the doctrine of *Allotropy*, or that the same elementary body is capable of assuming totally diverse forms, so different that it is difficult to identify them as the same substance. Thus, *sulphur* presents itself in three, if not four, modifications, which differ entirely in colour, in density, in melting point, in solubility, and in crystalline form.

208. Combinations.—*Sulphur* forms with *hydrogen* two compounds—

Sulphuretted hydrogen, $\mathbf{S}H_2$.
Hydrosulphyl, $\mathbf{S}'_2H_2$, or Hs_2.
With *carbon*, only one—
Carbonic disulphide, $\mathbf{C}S''_2$.

It combines rapidly with *chlorine*, *bromine*, and *iodine*, especially when the action is favoured by heat.

It also combines eagerly with nearly all the metals, not even excepting *silver*, *gold*, and *platinum*, forming with them *sulphides* or *sulphurets*. The sulphides of the metals are all of them insoluble, with the exception of those of Ba, Sr, Ca, Mg, K, Na, Am; and as they all possess a characteristic colour or other distinctive, this reaction with sulphur is of great importance in analysis.

Sulphur also combines with *oxygen* and *hydrogen* to form a series of two anhydrides, and no less than eight acids, the principal of which will be considered in the following chapter.

209. Uses.—*Sulphur* is extensively used in the arts in the manufacture of matches, on account of its inflammability. Large quantities are consumed in the manufacture of gunpowder. In the form of *sulphurous acid* it is used to bleach silks, flannels, and other substances, which would be destroyed by the action of *chlorine;* but its chief use is in the manufacture of *sulphuric acid*, so important both to the manufacturer and the chemist.

210. Sulphuretted Hydrogen, Hydro-sulphuric Acid, Hydric Sulphide.—Symbol, $\mathbf{S}H_2$, Ⓗ-Ⓢ-Ⓗ. Molecular weight = 34. Molecular volume, ☐☐. One litre weighs 17 criths. Solid at − 85·5° C. (− 122° F.); liquid under a pressure of 17 atmospheres, at 10° C.

211. Occurrence. — Evolved, associated with other things, from volcanoes; found also in certain mineral matters, and in waters containing organic matter and sulphates.

212. Preparation.—1. By burning *sulphur* in *hydrogen* direct union takes place—

$$2H_2 \quad + \quad S_2 \quad = \quad 2\mathbf{S}H_2.$$

2. It is generally prepared by the action of *hydrochloric* or *dilute sulphuric acid* on *sulphide of iron.*

The apparatus employed may be that used for the

generation of *hydrogen*, or similar to that in fig. 22 (but without the application of heat, which in this reaction is not needed). The gas should be made to pass through a wash bottle, with water, in order to free it from any particles of acid which may come over with it. If a solution of hydric sulphide is required for laboratory use, it may be passed into a vessel of water, as shown in the drawing; but, if the gas itself be wanted in order to examine its properties, it may be collected at the pneumatic trough over warm water, in which it is very slightly soluble. The reaction that takes place in each case may be represented by the following equations:—

$\mathbf{Fe}S''$	+	$2HCl$	=	$\mathbf{S}H_2$	+	$FeCl_2$.
Ferrous sulphide.		Hydrochloric acid.		Sulphuretted hydrogen.		Ferrous chloride.

$\mathbf{Fe}S''$	+	$\mathbf{S}O_2Ho_2$	=	$\mathbf{S}H_2$	+	SO_2Feo''.
Ferrous sulphide.		Sulphuric acid.		Sulphuretted hydrogen.		Ferrous sulphate.

3. If, however, the *sulphuretted hydrogen* be required absolutely pure, it is best obtained by the action of *hydrochloric acid* on *antimonious sulphide*. In this case heat is necessary, and the arrangement shown in fig. 56 should be adopted. The advantage of this method is that the gas ceases to be given off the moment the source of heat is taken away—

$\mathbf{Sb}_2S''_3$	+	$6HCl$	=	$3\mathbf{S}H_2$	+	$2\mathbf{Sb}Cl_3$.
Antimonious sulphide.		Hydrochloric acid.		Sulphuretted hydrogen.		Antimonious chloride.

213. Properties.—A transparent, colourless gas, having an almost disgusting odour of stale fish and rotten eggs. This odour is perhaps worse when largely diluted with air than in the pure gas itself. Even when largely diluted with air, this gas is highly poisonous, less than 1 per cent. being quite sufficient to kill small animals, such as birds. Hence the necessity of preparing it in the laboratory in a well ventilated cupboard. *Sulphuretted hydrogen* is very soluble in water; at 0° C. water dissolves nearly $4\frac{1}{2}$ times its volume of the gas, and at 15° C. $3\frac{1}{4}$ times its volume. The gas burns in air or *oxygen* with a pale blue flame, producing *sulphurous anhydride.*

Iodine and *bromine* act in the same way in decomposing the gas.

By one of these ways the gas may be decomposed and rendered innocuous if it has escaped into the atmosphere of a room. Put some *chloride of lime* on a plate and sprinkle with some dilute acid, *chlorine* will be set free, and will act on the *sulphuretted hydrogen* according to the above equation. Or put a few crystals of *iodine* on a hot saucer, the same effect will be produced.

214. Reactions.—*Sulphuretted hydrogen* acts on the hydrates and oxides of the metals, producing sulph-hydrates and sulphides—

OKH	+	**S**H_2	=	**S**KH	+	**O**H_2.
Potassic hydrate.		Sulphuretted hydrogen.		Potassic sulph-hydrate.		Water.

CuO	+	**S**H_2	=	CuS″	+	**O**H_2.
Cupric oxide.		Sulphuretted hydrogen.		Cupric sulphide.		Water.

215. Use.—A solution of this gas is most valuable in the laboratory, since, when mixed with a soluble salt of a metal, it produces in nearly all cases an insoluble sulphide of the metal; thus, with Zn, the sulphide is white; with *Mn*, flesh coloured; Sb, orange; As, light yellow; Cu, brownish black; Cd, yellow; Pb and Bi, black, &c.

216. Test.—Strips of bibulous paper, dipped in a solution of *plumbic acetate*, called lead paper, are turned black by the gas.

217. Carbonic Disulphide, Prisulphide of Carbon.—Symbol, **C**S″$_2$— (S)=(C)=(S). Molecular weight, 76; molecular volume, ⊞. One litre (as vapour) weighs 38 criths; specific gravity, 1·26. Boils at 45° C. (49° F.)

1. Prepared on a large scale by driving the vapour of sulphur over glowing coke or charcoal—

C	+	S_2	=	CS″$_2$.
				Carbonic disulphide.

2. By heating together *carbon* (charcoal) and *iron* or *copper pyrites*—

C	+	$2\mathbf{Fe}S_2''$	=	$\mathbf{C}S''_2$	+	$2\mathbf{Fe}S''_2$.
		Ferric disulphide.		Carbonic disulphide.		Ferrous sulphide.

218. Properties.—This is a liquid of a slight yellow colour, very volatile, and of an exceedingly offensive odour; its vapour is poisonous both to animal and vegetable life, and is also *remarkably* inflammable. It is very much heavier than water, in which it is not soluble, and consequently, it is usually kept under a layer of water. It is very soluble in ether and alcohol, and also in the oils. It is one of the best solvents for oils and fat, and is therefore frequently used for their extraction.

It also dissolves very freely *sulphur*, *iodine*, *bromine*, and *phosphorus* (except the amorphous or red variety).

Exp. 89.—Put into four test tubes a few drops of *carbon disulphide*. To the first put a little *sulphur* in powder, to the second a few grains of iodine, to the third a *very small* piece of common phosphorus, and to the fourth a little water. Note the solubility of the sulphur, phosphorus, and iodine, as also the beautiful colour of the iodine solution, and the insolubility of the carbonic disulphide itself in water.

It combines with the alkaline hydrates, forming sulphocarbonates, which are analogous to the carbonates, only sulphur has taken the place of oxygen. These compounds are very unstable—

$6\mathbf{O}KH$	+	$3\mathbf{C}S''_2$	=	$2\mathbf{C}S''Ks_2$	+	$\mathbf{C}OKo_2$	+	$3OH_2$.
Potassic hydrate.		Carbonic disulphide.		Potassic sulpho-carbonate.		Potassic carbonate.		Water.

CHAPTER XVII.

COMPOUNDS OF SULPHUR WITH OXYGEN AND HYDROXYL.

219. Sulphur forms with oxygen and hydroxyl a large series of compounds, in which it plays the part of either a dyad, tetrad, or hexad. Some of these compounds are of the greatest importance in a chemical point of view. They are—

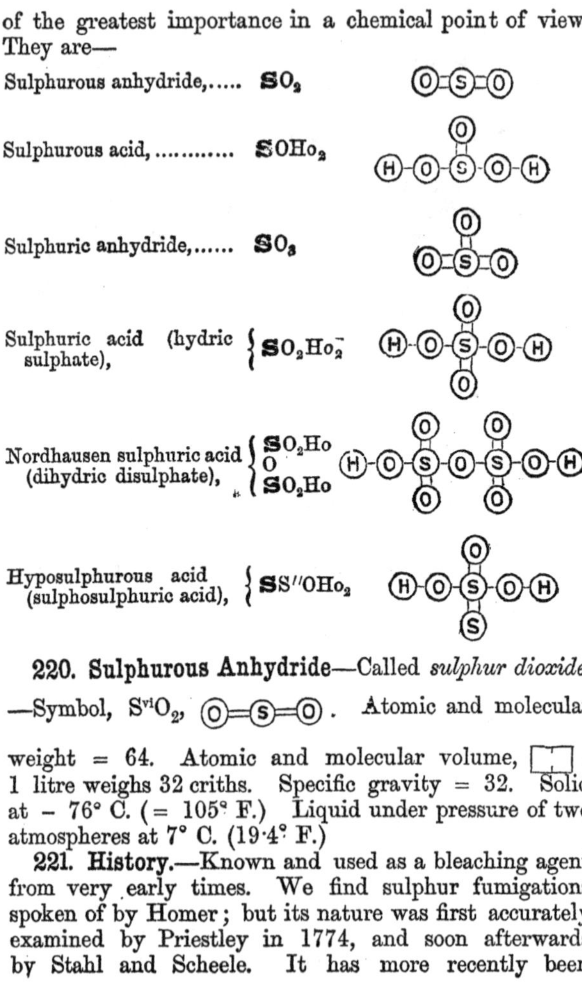

Sulphurous anhydride,..... $\mathbf{S}O_2$

Sulphurous acid, ... $\mathbf{S}OHo_2$

Sulphuric anhydride,...... $\mathbf{S}O_3$

Sulphuric acid (hydric sulphate), $\mathbf{S}O_2Ho_2$

Nordhausen sulphuric acid (dihydric disulphate), $\left\{\begin{array}{l}\mathbf{S}O_2Ho \\ O \\ \mathbf{S}O_2Ho\end{array}\right.$

Hyposulphurous acid (sulphosulphuric acid), $\mathbf{S}S''OHo_2$

220. Sulphurous Anhydride—Called *sulphur dioxide.* —Symbol, $S^{vi}O_2$, (O)=(S)=(O). Atomic and molecular weight = 64. Atomic and molecular volume, ⊔⊔. 1 litre weighs 32 criths. Specific gravity = 32. Solid at − 76° C. (= 105° F.) Liquid under pressure of two atmospheres at 7° C. (19·4° F.)

221. History.—Known and used as a bleaching agent from very early times. We find sulphur fumigations spoken of by Homer; but its nature was first accurately examined by Priestley in 1774, and soon afterwards by Stahl and Scheele. It has more recently been made the subject of investigation by Gay Lussac and Berzelius.

222. Occurrence.—Occurs naturally in the vicinity of volcanoes, where it escapes from the earth in a gaseous form. It is found also present in the air (as an impurity) in the neighbourhood of large towns, arising from the burning of the sulphur which is present to a greater or less extent in all coal. When it is required for use, however, it is always prepared artificially, for which there are several methods.

223. Preparation.—1. By the burning of *sulphur* in *air* or *oxygen.* Sulphur burns with a lilac flame and produces a permanent gas, which occupies the same space as the O, but is doubled in density; thus 2 volumes of O + 1 volume of S become condensed into 2 volumes of $\mathbf{S}O_2$—

$$S + O_2 = \mathbf{S}O_2 \text{ or } \boxed{O\,O} + \boxed{S} = \boxed{S\,O^2}$$

2. More generally prepared by heating metals as Cu or Hg with sulphuric acid; sulphurous anhydride is given off, water and cupric or mercuric sulphate being left behind. Thus—

$$2\mathbf{S}O_2Ho_2 + Cu = \mathbf{S}O_2 + \mathbf{S}O_2Cuo'' + 2\mathbf{O}H_2.$$

Sulphuric acid. Sulphurous anhydride. Cupric sulphate. Water.

The reaction with *mercury* would be precisely the same, Hg being substituted for Cu in the equation.

224. Properties.—At ordinary temperatures and pressures, a transparent, colourless gas, which has the peculiar odour of burning sulphur; it is neither combustible nor a supporter of combustion; it is very soluble in water, forming with one part of the water *sulphurous acid,* and then being dissolved in the remainder as a solution of *sulphurous* acid. This acid liquid when cooled to 0° C. deposits white cubical crystals of *sulphurous acid*—

$$\mathbf{S}O_2 + \mathbf{O}H_2 = \mathbf{S}OHo_2''.$$

Sulphurous anhydride. Water. Sulphurous acid.

This sulphurous acid possesses powerful bleaching properties, and may be kept unchanged for a long time if air be excluded; but if air gets access to it, it is rapidly converted into *sulphuric acid*, $\mathbf{S}O_2Ho_2$ [H_2SO_4]. It has a very great affinity for O, which enables it to bleach many vegetable colours, and thus makes it very important in the bleaching of all animal textures, such as flannel, sponge, silk goods, &c., which would be destroyed by *chlorine*. They are bleached by being exposed while damp in a closed chamber to the vapours of burning *sulphur*.

Sulphurous anhydride is also used as a fumigant to destroy insects, and to remove offensive odours, and so to destroy infection; it is used as a means to retard for a time putrefaction in meat, and to check fermentation in cider and home-made wines; for this latter purpose a little *sulphur* is burnt in the cask before filling it with the liquor. It is also used as a bath in some cases of skin disease.

In commerce it is very largely used to bleach straw hats and bonnets, &c., but its chief use is in the manufacture of *sulphuric acid*.

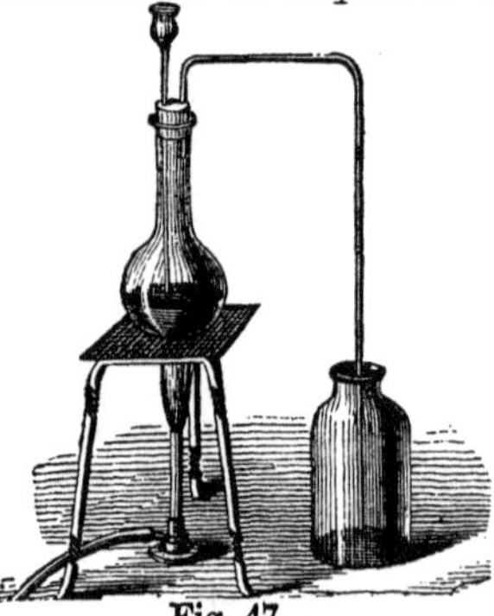

Fig. 47.

Exp. 90.—Place 5 grammes (80 grs.) of copper clippings in a flask with safety funnel and bent tube fitted through the cork; pour on 30 c. c. (1 oz.) of sulphuric acid, $\mathbf{S}O_2Ho_2$ [H_2SO_4], heat strongly, and collect the gas given off by downward displacement. (Fig. 47.)

Exp. 91.—Test the gas with a lighted taper; it will not burn itself, nor will it allow the taper to burn in it—*i. e.*, it is neither a combustible nor a supporter of combustion.

Exp. 92.—Test it with blue litmus paper. Observe the litmus paper is reddened, indicating the presence of a strong acid.

Exp. 93.—Suspend rose or other flower in jar of vapour of $\mathbf{S}O_2$; they will lose their colour. The colour, however, is not discharged as it would be by chlorine; for if the bleached flowers be thrown into a *very weak solution of ammonia* or *caustic soda* the colour will be first restored and then changed to green by the alkali. The action differs, too, on different colours; reds are bleached at once, while greens require a very prolonged action of the gas. Logwood also is bleached very slowly, while indigo is not at all affected.

EXP. 94.—Its great solubility is shown, as in the case of *hydrochloric acid*, by covering the mouth of a jar of the gas with a glass plate, inverting it, and removing the glass plate under water.

225. Reactions.—1. If *sulphurous anhydride* be passed into solutions of the metallic hydrates, salts known as *sulphites* are formed. Of these *sulphites* there are two—the *acid sulphite* and the *normal sulphite*, formed according as the *sulphurous anhydride* or the *metallic hydrate* is in excess.

226. Tests for Sulphites, and therefore for presence of Sulphurous Acid.—1. The *sulphites* are made known by their effervescing on the addition of any strong acid, as the *hydrochloric* or *sulphuric* acids, giving off a colourless gas with the pungent odour of *sulphurous anhydride.*

2. If a solution of any *sulphite* has a few drops of *baric chloride* ($BaCl_2$) added to it, a white precipitate of *baric sulphite* ($BaSO_3$) will be thrown down. If the original *sulphite* be pure and free from *sulphate*, this precipitate will be dissolved on the addition of *hydrochloric acid* (HCl). If now to the clear liquid some saturated *chlorine water* be added, it will immediately turn milky; the chlorine takes H_2 from the water to form HCl, and the O, set free, converts the *sulphurous acid* ($\mathbf{S}OHo_2$) into *sulphuric acid* ($\mathbf{S}O_2Ho_2$), which, uniting with the *barium*, forms *baric sulphate* ($\mathbf{S}O_2Bao''$), which is insoluble in all acids—

$\mathbf{S}OHo_2$	+	Cl_2	+	H_2O	=	$\mathbf{S}O_2Ho_2$	+	2HCl.
Sulphurous acid.		Chlorine.		Water.		Sulphuric acid.		Hydrochloric acid.

227. Sulphurous Acid.—Symbol, $\mathbf{S}OHo_2$. Molecular weight, 82. Crystallizes at 0° C.

This acid is formed whenever water or moisture in any form is brought into contact with *sulphurous anhydride*. It is very difficult to obtain this compound perfectly free from admixture with water.

The properties and reactions of *sulphurous acid* have been already described in treating of *sulphurous anhydride*—the latter, in fact, only acting through the intervention of moisture, whereby it becomes converted into *sulphurous acid*.

228. Sulphuric Anhydride.—Symbol, $\mathbf{S}^{vi}O_3$. Molecular weight = 80. Molecular volume, $\square$. One litre of sulphuric anhydride vapour weighs 40 criths. Fuses at 24·5° C. (76° F.) Boils at 52·6° C. (91·7° F.)

229. Preparation.—1. By passing a mixture of *sulphurous anhydride* and *oxygen* over spongy platinum ignited to a red heat. The platinum by a *catalytic* action induces the $\mathbf{S}O_2$ and O to unite, forming $\mathbf{S}O_3$, forming *sulphuric anhydride.*

2. By heating gently in a retort *Nordhausen* (or *fuming*) *sulphuric acid.* Fumes of *sulphuric anhydride* are given off, and may be condensed in a receiver, which is kept cool with ice—

$$\left\{\begin{matrix}{}'\mathbf{S}^{v}O_2Ho \\ O \\ {}'\mathbf{S}^{v}O_2Ho\end{matrix}\right. = \mathbf{S}^{vi}O_2Ho_2 + \mathbf{S}^{v}O_3.$$

Nordhausen sulphuric acid.	Sulphuric acid.	Sulphuric anhydride.

230. Properties.—*Sulphuric anhydride,* when condensed in a receiver kept cool by ice, assumes the form of a fibrous looking mass of white, silky crystals, very tough and ductile. It allows itself to be moulded by the fingers (if they are perfectly dry), without charring the skin. It fumes in the air, and is very deliquescent; and when mixed with water, it combines with such energy as to give rise to intense heat, and a hissing noise similar to that caused by the contact of red-hot iron with water. This solution has all the properties of *sulphuric acid. Sulphuric anhydride* possesses no acid properties; it will not redden litmus paper, nor in fact act in any way as an acid, until it has been brought in contact and combined with water, when, sulphuric acid having been formed, all the reactions due to that compound take place.

231. Nordhausen Sulphuric Acid—Symbol, $\left\{\begin{array}{l}\mathbf{S}O_2Ho\\ O\\ \mathbf{S}O_2Ho\end{array}\right.$ — is so called from its being chiefly made at Nordhausen, in Saxony. It may be regarded as ordinary *sulphuric acid*, having dissolved in it one proportion of *sulphuric anhydride.* This idea of its composition is clearly set forward in its atomic formula, (H_2SO_4, SO_3). The compound is so unstable, that, under ordinary circumstances of temperature and pressure, it fumes or gives off vapours of *sulphuric anhydride,* hence its name of fuming sulphuric acid.

232. Preparation.—Ferrous sulphate ($\mathbf{S}OHo_2Feo''$, $6\mathbf{O}H_2$) is dried at a moderate heat, to expel its water of crystallization, and is then distilled in earthen retorts; a dense, brown, fuming liquid passes over, having a specific gravity of about 1·9.

233. Uses.—The chief use of this *Nordhausen sulphuric acid* is for dissolving *indigo* for the preparation of that beautiful dye known as *Saxony blue.* Ordinary *sulphuric acid* will not dissolve it.

234. Sulphuric Acid.—Called also *hydric sulphate,* and *oil of vitriol.*—Symbol, $\mathbf{S}\ O_2Ho_2$ $[H_2SO_4]$. Molecular weight and combining proportion, 98. Molecular volume, ⊞. One litre of sulphuric acid vapour weighs 24·5 criths. Specific gravity of liquid, 1·85. Boiling point, 325° C. [617° F.]

235. History.—The process for making *sulphuric acia* or *oil of vitriol* from *sulphur* and *nitre* was first described by Basil Valentine, born in the beginning of the fifteenth century. It was first introduced into England in 1720.

236. Preparation.—*Sulphuric acid* may be prepared

in several ways as a laboratory experiment, by the action of *hydroxyl* on *sulphurous anhydride, oxygen* on *sulphurous acid*, or *water* on *sulphuric anhydride*, as shown in the three following equations :—

(1.) $\mathbf{S}O_2$ + Ho_2 = $\mathbf{S}O_2Ho_2$.
Sulphurous anhydride. Hydroxyl. Sulphuric acid.

(2.) $\mathbf{S}oHo_2$ + O = $\mathbf{S}O_2Ho_2$.
Sulphurous acid.

(3.) $\mathbf{S}O_3$ + $\mathbf{S}H_2$ = $\mathbf{S}O_2Ho_2$.
Sulphurous anhydride. Water.

237. Manufacture.—This acid, which is one of the most important, not only to the chemist, but also to the manufacturer, is made on a very large scale by passing *sulphurous anhydride* ($\mathbf{S}O_2$), vapours of *nitric acid* ($\mathbf{N}O_2Ho$), *steam* ($\mathbf{O}H_2$), and air, into a large leaden chamber, and allowing them to mix freely.

The *sulphurous anhydride* is obtained by burning sulphur in furnaces, the chimneys of which communicate with the leaden chamber. Above, and in the flame of each furnace, is hung an iron pot charged with a mixture of *sodic* or *potassic nitrate* and *sulphuric acid*, which under the influence of heat becomes converted into *sodic sulphate* and *nitric acid*, thus—

$2\mathbf{N}^{v}O_2Nao$ + $\mathbf{S}^{vi}O_2Ho_2$ = $\mathbf{S}O_2Nao_2$ + $2\mathbf{N}^{v}O_2Ho$.
Sodic nitrate. Sulphuric acid. Sodic sulphate. Nitric acid.

The vapours of the *nitric acid* thus enter the chamber in company with those of *sulphurous anhydride.*

The steam, $\mathbf{O}H_2$, is supplied from a boiler which has a series of jets opening into the chamber; while the air is obtained by a strong draught, caused by a tall chimney at the end of the chamber opposite the furnace. This tall chimney also serves for the escape of any excess of the gases, which, in a well managed chamber should consist only of *nitrogen* and *nitric oxide*, the *sulphurous anhydride* and *oxygen* being supplied in just sufficient quantities to condense each other.

The changes which take place when these four substances meet each other are as follows :—The *sulphurous acid* (formed by the union of the *sulphurous anhydride* and *steam*) has the power of deoxidizing the *nitric acid*, by depriving it of oxygen, and converting it into *nitric oxide*, becoming itself changed into *sulphuric acid*.

$$2\mathbf{N}o_2Ho \ + \ 3\mathbf{S}O_2 \ + \ 2\mathbf{O}H_2 \ = \ 2\mathbf{N}O \ + \ 3SO_2Ho_2.$$

When *nitric oxide*, $\mathbf{N}_2O_2$, comes in contact with free O, it becomes converted into *nitrous anhydride* ($\mathbf{N}_2O_3$). This it does in the leaden chamber obtaining O from the air, thus—

$$\mathbf{N}_2O_2 \ + \ O \ = \ N_2O_3.$$

Nitric oxide. Nitric anhydride.

This N_2O_3, in the presence of *sulphurous anhydride* and *water* (steam), parts with an atom of O to form sulphurous acid, being itself reduced to N_2O_2, which again takes O from the air, and so on—

$$\mathbf{N}O_3 \ + \ \mathbf{O}H_2 \ + \ \mathbf{N}O_2 \ = \ \mathbf{N}_2O_2 \ + \ \mathbf{S}O_2 \ Ho_2.$$

Nitrous anhydride.		Sulphurous anhydride.	Nitric oxide.	Sulphuric acid.

The sulphuric acid thus formed is immediately dissolved in the water which covers the bottom of the tank to the depth of 3 or 4 inches. The liquid collected at the bottom of the chamber is very dilute *sulphuric acid;* it is then drawn off and concentrated by evaporation. This is carried on in leaden pans, at a gentle heat, until it attains a specific gravity of 1·72, when it begins to attack the lead. At this stage it is of a brown colour, and is known as the *brown oil of vitriol.* It is finally concentrated in vessels of glass (bottles holding about 40 gallons), or else in boilers made of platinum. One of these platinum boilers was exhibited at the British Association Meeting at Nottingham, which cost £6,000.

The *nitric acid* supplies the first amount of oxygen to convert the *sulphurous* into *sulphuric acid*, being itself reduced to *nitric oxide*, which thenceforth serves as a mere carrier of *oxygen*, absorbing it from the air, and parting with it again to the *sulphurous acid.* It would be unnecessary to add fresh quantities of these carriers (the products of the *sodic nitrate*) were it not that some of it is continually escaping into the tall chimney with which the chamber is connected, and other parts of the *nitrous anhydride* become converted by reduction into *nitrous oxide.*

In distilling over the brown acid of commerce, care must be taken that all traces of nitrous compounds must be removed before employing platinum vessels, as otherwise they become corroded. The water which comes over has always a small proportion of sulphuric acid with it; it is therefore preserved and returned to the leaden chamber.

238. Properties.—*Sulphuric acid* (the oil of vitriol) of commerce is, when pure, a colourless, transparent, and odourless liquid, dense and oily-looking (hence its name), and of a specific gravity 1·842. It is of an intensely acid and caustic nature, and, on account of its great affinity for water, chars all organic substances (as wood, sugar, &c.); this it does by removing O and H in the form of $\mathbf{O}H_2$, and setting free the carbon, C. Its affinity for water is so great that, if left exposed in an open vessel for two or three days, sometimes even in twenty-four hours, it will double itself in weight from the aqueous vapour absorbed from the air. When the acid and water are mixed, great heat is given out; the dilution should always be made in *thin* glass vessels, and by pouring the acid gradually into the water and stirring, and *not* the water into the acid. This property of absorbing water makes it very valuable in the laboratory as a means of drying gases. That the action between the sulphuric acid and water is a chemical one, and not a mere dilution, is shown by the following facts:—1st, That heat is produced; 2nd, that condensation takes place; and, 3rd, that the water cannot be entirely separated by evaporation.

239. Reactions.—*Sulphuric acid* is dibasic—that is to say, it forms two principal classes of salts with monad metals exemplified in—

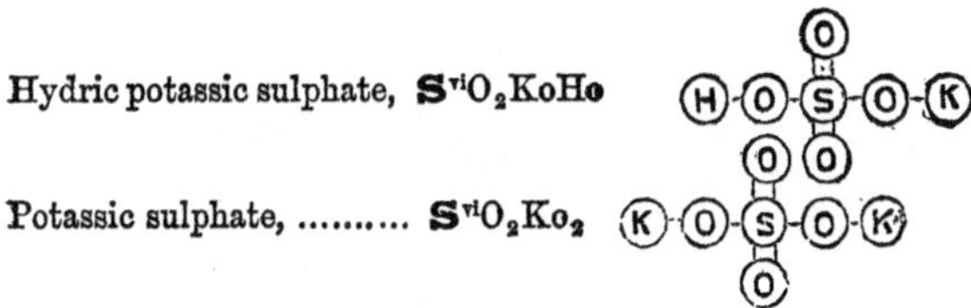

The action of this acid on metals varies; when poured cold and pure on some metals, such as *zinc* and *iron*, a very slight action takes place, which is almost immediately stopped, the reason being that a salt (a sulphate of the metal) is formed which protects the remainder of the metal from further change; but if a little water be added, the action is immediately quickened, the metal is dissolved, and *hydrogen* is evolved. The reason is, that the salt is dissolved by the water as soon as formed, and a fresh surface of the metal is exposed to the action of the acid.

On such metals as *copper, mercury, arsenic, tin, lead, silver*, &c., *sulphuric acid*, even when concentrated, has no action when cold; but when boiled with them, *sulphurous anhydride* and the sulphate of the metal are formed—

$$2\mathbf{S}^{vi}O_2Ho_2 + Ag_2 = \mathbf{S}^{v}O_2Ago_2 + 2\mathbf{O}H_2 + \mathbf{S}^{vi}O_2.$$

Gold and *platinum* undergo no change even when boiled with *sulphuric acid.*

The manufacture of sulphuric acid is shown in Fig. 46, in which A A represent furnaces in which *sulphur* or *iron pyrites* is burned; in which an iron pot is suspended, charged with *sodic nitrate* and *sulphuric acid.* Vapours of nitric acid are thus liberated, which pass on along with those of sulphurous anhydride into immense leaden chambers, F F, supported by strong timber framework. Steam is projected from a boiler, E, through a pipe C, into the chamber. This not only supplies the moisture necessary for the first stage of the operation, but also facilitates the mixing of the various gases.

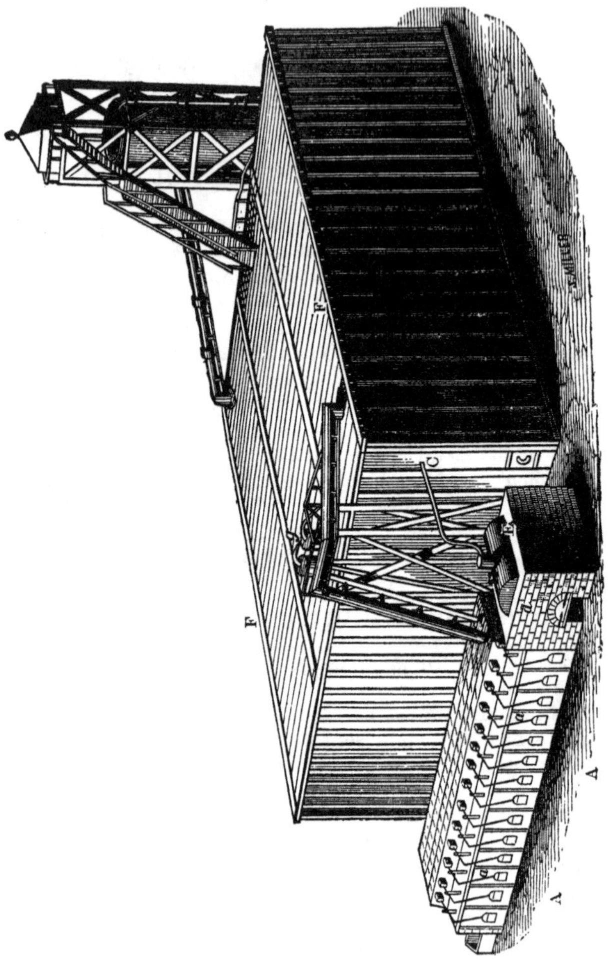

Fig. 46.

Some idea of the extent of these chambers may be formed from their size, from 4 to 5 metres (12 to 15 feet) high, 5 to 7 metres (14 to 21 feet) wide, and 50 to 100 metres

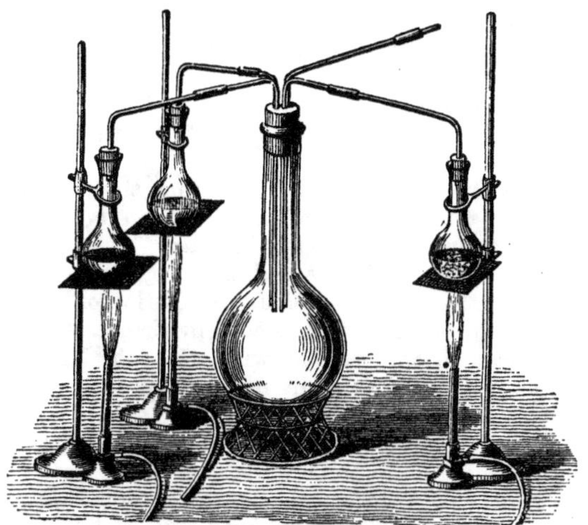

Fig. 47.

(150 to 300 feet) in length. They are sometimes intersected by partial leaden partitions to effect the mixture of the gases more perfectly.

The reaction is easily seen on a small scale in the laboratory or at the lecture table, by means of the apparatus, Fig. 47, in which a large globular receiver has four bent tubes fitted through its cock, three of which are connected with flasks, giving off respectively *sulphurous anhydride, nitric acid,* and *steam;* while the fourth supplies *air* or *oxygen.* If no water be present in the receiver, in a few minutes its inner surface becomes cooled, with a white crystalline deposit, into which *sulphurous anhydride, nitric peroxide,* and *oxygen* enter. As soon as this mass is treated with water it is decomposed with effervescence, *sulphuric acid* being formed, and *nitrous anhydride* set free.

240. Impurities.—The chief impurities in *sulphuric acid* are *plumbic sulphate*, formed by the slow action of the acid on the lead pans in which it is evaporated; fumes of *nitrogen* compounds dissolved in the acid, and derived from the nitric acid employed in its manufacture; and sometimes *arsenic*, when it has been prepared from *iron pyrites*. The mode of detecting each is as follows:—

The *lead* may be detected by diluting the acid with water, when the sulphate falls in the form of a white powder, being insoluble in *cold dilute acid*.

The compounds of N may be detected by boiling some of the acid with a few drops of dilute solution of *indigo*, the blue colour of which is destroyed by them.

The *arsenic* is only present when the acid has been made from *pyrites;* it is easily detected by adding a few drops of *hydrogen sulphide*, when the *arsenic* is thrown down, as an insoluble yellow sulphide.

241. Tests.—1. Any soluble salt of *barium* gives, with *sulphuric acid*, a white precipitate of *barium sulphate*, $\mathbf{S}O_2Bao_{,}'$ [$BaSO_4$], which is insoluble in all acids. This precipitate takes place even when the acid is very dilute.

2. The strong *acid* produces great heat when poured in water.

3. The strong *acid* blackens wood, sugar, &c.

242. Uses.—The applications of *sulphuric acid* in the arts are almost innumerable, while to the chemist it is perhaps more important and essential than any other substance with which we are acquainted. The brown acid of commerce is used, without purification, in the manufacture of the so-called "*superphospate of lime*," &c., for manure. Very large quantities are consumed in the manufacture of *sodic sulphate*, which is a preliminary process to the formation of *sodic carbonate*. It is also used in large quantities in the preparation of *nitric* *hydrochloric* and other volatile acids, besides many other important compounds. Its uses in the laboratory are too numerous to mention; in fact, it forms one of the most important reagents we have.

Its most important use is perhaps the manufacture of soda from common salt. On this production of soda depend the two manufactures of soda and glass. It has also materially simplified and cheapened the refining of silver, and the separation of gold, which is always present; and in the manufacture of artificial manure (*superphosphate of lime*) from bones it is almost invaluable.

The discovery of a cheap and easy method of making sulphuric acid has already formed one epoch in chemistry; and every fresh discovery which either simplifies or cheapens the production of this important compound will form a starting point for a new era in the science.

INDEX.

ACID, definition, 46.
  Frankland's definition, 46.
Affinity, characters, 18.
Alkali, definition, 48.
Alloys, 12.
Amalgams, 12.
Ammonia, 157.
  Analysis, 161.
  History, 157.
  Occurrence, 157.
  Preparation, 157.
  Properties, 158.
  Synthesis, 160.
  Test, 160.
Ammonic Salts, 201.
Ammonium, 161.
Artiads, 58.
Atmosphere, Composition, 148.
Atomic theory, 31.
Atomicity, definition, 54.
  Absolute, active, and latent, 59.
  Table of, 56.

BASE, definition, 48.
Boric, boracic, or orthoboric acid, 131.
  Occurrence, 131.
  Properties, 132.
Boric anhydride, 131.
  Preparation, 131.
Boron, 129.
  Allotropic modifications, 129.
  History, 129.
  Occurrence, 129.
  Preparation, 130.
  Properties, 130.
Boyle and Marriotte, law of, 36.
Brackets, effect of, 51.
  Use in graphic notation, 64.

CARBON, 133.
  Allotropic forms, 143.
  Amorphous, 144.
  Combinations, 146.
  Occurrence, 133.
  Graphitic, preparation, 134.
  Graphitic, properties, 134.
  Graphitic, uses, 134.
Carbonic anhydride, 136.
  Analysis, 142.
Carbonic anhydride—
  Formation, 137.
  History, 137.
  Occurrence, 137.
  Preparation, 137.
  Properties, 138.
  Synonyms, 137.
  Synthesis of, 141.
Carbonic disulphide, 168.
  Properties, 169.
Carbonic oxide, 143.
  Formation, 143.
  History, 143.
  Preparation, 143.
  Properties, 144.
Charcoal, animal, 134.
  Wood, 134.
  Properties, 135.
Chemical attraction, characters of, 18.
  Summary of, 25.
  Chemical combination, laws of, 26.
  Nomenclature, principle of, 43.
Chemistry, definition, 11.
Chlorates, preparation and properties, 127.
Chloric acid, 126.
  Preparation, 126.
  Properties, 127.
Chloric peroxide, 124.
  Preparation, 124.
  Properties, 126.
Chlorine, 82.
  Combinations, 89.
  Distribution and natural history, 82.
  History, 82.
  Preparation, 83.
  Properties, 85.
  Synonyms, 82.
Chlorous anhydride, 123.
  Preparation, 123.
Chlorous acid, 126.
  Preparation, 126.
Coke, 136.
Combining volumes, 34.
Combining weights, 26.
Compound bodies, definition, 11.
Crith, uses of, 41.

DIFFUSION, 75, 78.

Dyads, 55.

ELEMENTS, basylous and chlorous, 44.
 Definition, 11.
 Distribution, 15.
 List of, 13.
 Order of positive and negative, 45.
 Physical condition of, 15.
 Relative importance, 15.
Equations, chemical, 52.
Equivalence, definition, 54.

FORMULÆ, chemical, 50.
 Rational and empirical, 63.
Forces, 9.

GASES, correction for temperature and pressure, 37.
Graphic notation, 57.

HEXADS, 55.
Hydrogen, arithmetical considerations, 80.
 As a metal, 81.
 Combinations, 79.
 Demonstration of lightness, 77.
 Distribution, 66.
 History, 66.
 Preparation, 66.
 Properties, 74.
Hydrochloric acid, 90.
 Analysis, 97.
 Distribution, 91.
 Impurities, 95.
 Preparation, 91.
 Properties, 92.
 Reactions, 98.
 Solution, 94.
 Synonyms, 90.
 Tests, 95.
 Uses, 98.
Hydroxyl, 119.
 Preparation, 119.
 Properties, 120.
 Reactions, 120.
Hypochlorous acid, 124.
 Preparation, 124.
 Properties, 125.
Hypochlorous anhydride, 121.
 Preparation, 122.

ISOMERISM, 27.

LAMPBLACK, 136.
Laughing gas, 152.
Law of Boyle and Marriotte, 36.

MATTER, definition, 9.
 Simple and compound, 11.
Mechanical mixture, properties, 16.
Metals, characteristics of, 12.
Monads, 55.

NITRIC acid, 149.
 Manufacture, 150.
 Preparation, 150.
 Properties, 151.
 Tests, 151.
Nitric anhydride, 152.
 Preparation, 152.
Nitric oxide, 153.
 Preparation, 154.
 Properties, 154.
Nitric peroxide, 156.
 Preparation, 156.
Nitrogen, 145.
 Combinations with oxygen and hydroxyl, 148.
 History, 145.
 Occurrence, 145.
 Preparation, 145.
 Properties, 147.
Nitrous Acid, 155.
 Preparation, 155.
Nitrous anhydride, 155.
 Preparation, 155.
 Properties, 155.
Nitrous oxide, or nitrogen protoxide, 152.
 Preparation, 152.
 Properties, 152.
Nomenclature of acids, 47.
Nomenclature of compounds, 45.
Non-metals, characteristics, 12.
 List of, 15.
Notation, symbolic, 50.

OXYGEN, 98.
 Arithmetical considerations, 107.
 Distribution and natural history, 98.
 History, 98.
 Preparation, 98.
 Properties, 103.
 Synonyms, 98.
 Tests for, 103.
Ozone, 109.
 History, 109.
 Preparation of, 109.
 Properties, 110.
 Tests for, 111.

PENTADS, 55.
Perchloric acid, 129.
Perissads, 58.

QUANTIVALENCE, definition, 54.

RADICAL, simple, 61.
 Compound, 61.

SALTS, definition, 49.
Haloid, 49.
Haloid, nomenclature, 50.
Soot, 136.
Sulphosalts, 49.
Sulphur, 163.
Allotropic modifications, 165.
Combinations, 165.
Compounds with oxygen and hydroxyl, 169.
Extraction, 164.
History, 163.
Occurrence, 164.
Properties, 164.
Uses, 166.
Sulphites, tests for, 173.
Sulphuretted hydrogen, 166.
Occurrence, 166.
Preparation, 166.
Properties, 167.
Reactions, 168.
Test, 168.
Use, 168.
Sulphuric acid, 175.
History, 175.
Impurities, 182.
Manufacture, 176.
Preparation, 175.
Properties, 178.
Tests for, 182.
Uses, 182.
Nordhausen, 175.
Preparation, 175.
Sulphuric anhydride, 174.
Preparation, 174.
Properties, 174.
Sulphurous acid, 173.
Tests for, 173.
Sulphurous anhydride, 170.
History, 170.
Occurrence, 171.
Preparation, 171.
Properties, 171.
Reactions, 173.
Symbols, 13.

TETRADS, 55.
Thick type, use of, 63.
Triads, 55.

VOLUME weights, 35.

WATER, 112.
Water, composition, 113.
Electrolysis, 73.
History, 112.
Preparation, 112.
Properties, 115.
Pure, 115.
Reactions, 116.
Tests for impurities, 118.
Weight, 38.
French system, 39.
French system, conversion to English, 40.

WILLIAM COLLINS, AND CO., PRINTERS, GLASGOW.

www.ingramcontent.com/pod-product-compliance
Lightning Source LLC
LaVergne TN
LVHW011225110826
845150LV00006B/1550